Genetic Engineering

1/8/97

Sent with pleasure
to Dr. Glenn
Mcgee —

Theodore Kent

GENETIC ENGINEERING, YES, NO OR, MAYBE?
A Look At What's Ahead

Theodore C. Kent, Ph. D.

BookWorld Press, Inc.
Sarasota, Florida

Produced by BookWorld Press, Inc.

Distributed by
BookWorld Services, Inc.
1933 Whitfield Park Loop
Sarasota, FL. 34243
24 hour order line: 1-800-444-2524

Library of Congress Catalog Card Number 96-84461

Publisher's Cataloging in Publication

Kent, Theodore C., 1948–
    Genetic Engineering : yes, no, or maybe : a look at what's ahead
/ Theodore C. Kent
    p. cm.
    Includes bibliographical references and indexes.

    ISBN: 1-884962-05-X

    1. Genetic engineering—Moral and ethical aspects. 2. Human
reproductive technology—Moral and ethical aspects. I. Title.

QH442.K46 1997   174'.25
                                  QBI96-40261

To *Home sapiens*—an endangered species.

TABLE OF CONTENTS

PREFACE

Almost weekly, our newspapers and popular magazines describe progress in genetic engineering. It is not unusual to read that another new gene has been discovered that contributes to causing a disease or one that gives crops a longer shelf-life or helps them to resist insect infestation. In spite of this good news, our growing ability to redesign life, including human life, remains controversial and even frightening.

Some people are convinced that all is safe. They trust that we shall know where to stop in rearranging life's building blocks and they have faith in government and private supervision of human genetic engineering. Others are not so sure. When considering whether or not we should continue to pursue genetic engineering, many religious leaders and some behavioral scientists, like myself, take a long look ahead and think yes, no, or maybe.

At this time, we do not yet have the ability to redesign human nature by means of gene technology. Nevertheless, we have reached the point in molecular research where we should question where it may lead us. We have begun the journey and must now explore the road ahead of us. In doing so, we should keep in mind that, like other innovations in science, molecular technology's progress will be propelled forward by its self-generated momentum. A time could come when this momentum may sweep us along with it, since it promises the birth of new industries and offers those who control gene technology vast wealth and power.

The thought of changing the genetic blueprint of our species is disturbing to some of us who look far into the future. Inevitably, manipulating the human genome raises questions of defining human life, of morality, and even of the nature of reality. It raises fears that if, in future years, gene technology is applied to eliminate destructive human inherited tendencies it might get out of hand and deprive our spiritually-endowed, intellectually-gifted human species of free will. Is this an alarmist's fear or does it have substance? In the first two chapters, I argue that there is a real basis for fear for our human genome—the molecule we have in every one of our cells that contains our DNA.

The content of this molecule well might represent the universe's greatest treasure. Some biologists have called our attention to the awesome fact that the cells in each of our human brains are capable of making more interconnections than there are atoms in the entire universe. Is it prudent to experiment with such a marvelous structure even if we only intend to modify human traits that society views as destructive?

The lure of redesigning ourselves may be inherent in our human nature. Many are not aware that genetic engineering belongs to a category of human behavior that reaches back into antiquity. Since ancient times people have cherished the goals of redesigning themselves and have carried out forerunners of genetic engineering in large and small ways. People change clothes, style their hair, shave, wear rings and decorate themselves in various ways—all with the intentions of altering their natural appearance. We arouse our emotions artificially when we drink alcohol, respond to music, dance and listen to the beat of drums. All of these activities could be viewed as early forms of engineering ourselves. Therefore, genetic engineering travels a well worn road. It merely offers a new and more efficient way to achieve what we have always wanted.

I make the point in these pages that we are not yet emotionally ready or ethically prepared for the role that, in times ahead, genetic engineering may play. Before we permit gene technology to alter our lives, we must first make important environmental alterations to prepare ourselves for its application. Thereafter—cautiously—we may consider replacing a prudent "no" to further research involving some human inherited characteristics with a limited "yes." The goal of this book is to explore the road we must take on this journey.

Theodore C. Kent

Acknowledgments

My wife, Shirley, helped edit the book and made valuable suggestions. I used her as my consultant in writing this book, as I have done with previous books.

I appreciate the support of friends and colleagues who encouraged me to write a second book on the consequences of genetic engineering. Thanks are due to Paul A. Bergin and Patricia Middleton for editing the manuscript and arranging for the book's cover. I appreciate the work of Christopher Robinson, general manager of BookWorld Press, who was responsible for getting the book into final production. I am indebted to Robert A. Steel for helping with compiling the Index. I thank the University Press of America for granting me permission to incorporate some portions of my previous book *Mapping The Human Genome* into this book.

A part of Chapter Fourteen was adapted from my article in the *Truth Seeker* (vol. 122, No. 4, 1995) and quoted with permission.

INTRODUCTION

Ahoy, you who set sail in fragile barque!
In eagerness to hear my message,
You follow in wake of my sail,
Humming as it plows the sea.
Look back and reflect upon
Your own familiar shores.
Do not commit yourselves
To sail the seas I sail.
Lest, loosing sight of me, you founder.
The uncharted waters I traverse,
Have never been crossed before.

Thus does Dante warn us to think twice before we enter Paradise. There is a new paradise being offered us in the form of a panacea designed to remedy all our ills and even, eventually, to transcend our limitations. It offers us the hope and promise that we will, in the future, be able to create the "new human being" and the "new world" in the image of our choice. Indeed, the stunning successes of modern biological sciences in medicine, industry, agriculture, and especially genetics, has made us very optimistic about the future possibilities of realizing this dream. *But at what cost?* What are the implications of pursuing this dream? And if we attain it, what is it about ourselves as human beings that we risk losing?

The chapters that follow are, among other things, an attempt to get us to think now about such important issues. Mankind has always had utopian aspirations. We are driven by the desire to solve, once-and-for-all, the serious problems in life and to create our own Eden. And our latest version of the "original temptation" comes in this form: if only we can let the genome out of the bottle, we will indeed "be like gods" (Genesis 3:5).

But even if this were possible, would it be desirable? What would it mean in regards to the richness of human diversity, with our being creatively in the world, and, in turn, being "created" by the world we are in? The Spanish philosopher José Ortega y Gasset reminds us that we human beings are, above all, future-making and at the same time, aswarm with hopes and fears. This observation by a man who was preoccupied with the question of what it means to be human, underscores a *leitmotif* that runs throughout this book. Namely that of *Homo habilis,* a can-do ancestor of mankind who activates, by his behavior, the inherent potentialities of the universe. However, in understanding this concept, one must avoid succumbing to the notion common in Western traditional philosophy, that the distinctive human capacities for abstraction and system building which we designate by the concept "mind," function merely as a "mirror of Nature" manifesting what is already there, albeit in novel ways.

It is almost as if all we really do is *explicate* rather than truly *create* by our activation and further potentiating these possibilities. Nor does being creatively involved in the world mean that we do, or can, tame Nature to obey our dictates. Creative "orderers" though we be, we are part of an interactive system. As we form, we are in turn formed.

Aristotle stated that all philosophy begins in wonder. It would end in wonder, too, if it were not for "Nature's miracle," the creative imagination. Why? Because this species specific (only we humans have it) ability enables us to see not only what is there, but something new everywhere, if only as a possibility! It is a thinking "which builds as it discovers" and will continue to build as long as we are willing to engage in the "loving struggle" it entails. As Einstein observed, imagination is more important than intellect.

Struggle. This is another theme that is woven throughout this book. What is its significance in the context of our concerns? Most importantly, because it is the antidote for our desire to be in "peaceful possession," as theologians were wont to say, of indisputable truths and/or absolute certainties. In reality, this desire is another form of Nietzsche's will to power and is a far cry from the genuine will to truth espoused by Jesus. This latter is an on-going *doing,* a continual self-questioning; in sum, a sense of humility. It eschews the arrogance of power that asserts we can once-and-for-all possess the "knowledge of good and evil" and thus become lords of the universe. To avoid this "original sin" of *hubris*, we need to become comfortable with the discomfort of keeping ourselves "off balance," to exist in the questioning mode. Otherwise, we risk losing that restlessness which enables us to continue to wonder and to exercise our creative imagination which "displays before us the ever changing picture of the possible" and its relatedness. Thus, the need for thinking in terms of the Big Picture.

Why is this an especially important perspective? Because, unless we embrace it, we risk falling victim to the detrimental effects of *fragmentation*. This is the tendency to take the part for the whole, to enclose ourselves in our own limited, narrow-scoped explanatory system and rest content that we have asked the right questions and given the correct answers, or are at least on the way to that goal. An example is the claim that knowing the complete human genome we will know what it is to be human. Indeed, we will have reached a state of "eternal rest," of *inertia!* And so we will have reached the end of our struggles.

How tempting. But is it worth it? I hold, along with the author, that the "unstruggling life is not worth living" (*Pace*, Socrates). Why? Because it precludes us from thinking in terms of the Big Picture, from *transcending*. Resting content in all-encompassing explanations, we will disengage from the effort to continually broaden our horizons, to embrace as over-arching a view as possible of the interrelatedness of our universe. In a word, we will lose something distinctively human.

Not to be seduced by this temptation demands that we activate our potential for magnanimity, our ability to be great-of-soul enough to undertake the struggle involved. In addition, it demands that we abandon what Paul Zweig has called "the sociology of Narcissus" and embrace willingly the sociology of community, that sense of oneness with each other, both locally and globally, and with our ecosystem. Consequently, it means we must not let our can-do nature result in the exploitation of our accomplishments without any consideration of the long-term consequences of our newly acquired and rapidly expanding technological powers. Essentially, this is a moral struggle and, to achieve this vision, we will be required to develop a much broader, more encompassing moral framework. This will not be an easy task.

The ethical issues raised by modern scientific research, especially in the biological sciences, are unprecedented. And complex as they are, they will become even more complex as this research proceeds. As we struggle with the inevitable ethical dilemmas that will arise, we must keep in mind that the process of solving them is not a simple linear one. As this book makes abundantly clear, we exist in a relationship of reciprocity with the universe. We do not stand apart, nor can we, from this system of mutual input/feedback/output. Therefore, any attempt at devising a new moral framework, must be guided by the nature of this relationship. Given all the factors and, above all, the institutional factions involved, this is a daunting task. We might consider it analogous to a "Copernican Revolution," a breaking apart of the perfectly ordered, hierarchically structured, totally predictable Aristotelian universe. Comforting though that was, it was unreal. The revolution came about because those who initiated it dared to ask different kinds of questions, to pursue different kinds of goals. And just as in the past, we can expect that any attempts to devise a new framework will involve a struggle with the entrenched institutions of moral authority. To work this through with some degree of success, we will, now more than ever, have to rely upon our creative imagination. Only thus can we confront the dilemmas and choices that arise from our creative interventions.

It is just such issues and questions that Dr. Kent's book addresses. I can only conclude by saying *"Tolle et lege."* Pick it up and read it. But beware! *Human Engineering, Yes, No or Maybe?* may prod you to undertake the arduous task of undergoing a major conversion.

Francis J. Marcolongo, Ph.D.

CHAPTER ONE

GENETIC ENGINEERING? NO.

The observations of modern history make it clear that the driving force behind the majority of contemporary scientific research throughout the world is competition for power, wealth, and reputation. To round out the picture, we must add to these the motivation created by challenge and curiosity. In themselves, these driving forces may not be altogether undesirable. Human nature may require them to get things done. In the case of research that will create industries that could alter the basic mechanisms of life, these goals, as the major incentives, are insufficient. The reason is that manipulating the basic stuff of life by means of genetic engineering differs fundamentally from any previous enterprise in human history. Therefore, it may lead to unforeseen results.

Genetic engineering holds out great promise for ending inherited abnormalities and increasing our food supply. Nevertheless, the possibility exists that, over extended time, our lack of a commitment to the special kind of ethics that gene technology requires may cause genetic engineering to do more harm than good. If this sounds unnecessarily pessimistic, I ask the readers to withhold judgment until after they have read the first two chapters of this book. Genetic research will build huge new industries and the possible consequences will create a future that is now pressing itself upon us. What we do with the basic building blocks of life in our biological laboratories will determine how our children and grandchildren will live. All around us there are signs that our modern ethics are not up to this task.

Surprisingly, this view of gene technology could be an optimistic one for humankind. Our discovery of the basic secrets of life may force us to adopt a morality to which most people today give only lip-service. If the threat of misuse of gene technology provides us with a new moral outlook, this result will outweigh in significance all that we may achieve by using genetic engineering. The danger that genetic engineering presents lies not as much in itself as in its mix with the norms of our contemporary culture. We must say no to genetic engineering because we live and are shaped by this culture and its present priorities.

It is now time to ask, will our human genome—the molecule that contains our DNA with its approximately 100,000 genes—be safe in our world today? The thoughtful answer can only be no. More than a half a century ago our world was described by Karl Menninger (1893–1966), the co-founder of the Menninger Clinic. At the mid-century he wrote that "no one will maintain today that all's right with the world. It is full of hate and murder and hunger and bitterness. There is waste and pessimism and fear and sorrow."

The world has not improved since Dr. Menninger's time. The question that naturally follows is: Will humankind ever be ready for the advent of gene technology? We shall explore this question further along in this book. I shall present my case to show that, in time, genetic engineering is likely to become open-ended, permitting human beings to engineer themselves and each other with only a minimum of restraint. Let us accept this hypothesis for the moment. Later, I shall give it the weight of further thinking on the subject. We have much to lose. Therefore, it may be time now to stop to consider the possible future consequences of what we are doing now.

If we adopt genetic innovations before environmental ones, we shall be putting the cart before the horse. We must rely on our present selves, not as yet genetically engineered, to bring about a way of life that will deter misuse of gene technology in the future. Under present circumstances it would be prudent to say no to further steps taken to rearrange the basic building blocks of life—including human life. We are still inclined to see the small, close-up picture of gene technology and not the large one we should be looking at if we wish to insure our safety.

Modern breakthroughs in electronics, atomic energy, rocketry, and medicine highlight our impressive human ingenuity. On the other hand, some of these scientific achievements carry the risk of pollution and misapplication. More worrisome is the fact that our greatest scientific discoveries become dangerous in the hands of those who make misleading claims about them. Few things could equal the potential for danger than those posed by misleading claims for various redesigned forms of life. No one is more aware of this than the members of the bioethics groups that are now scattered throughout this country and abroad. Religious leaders and concerned citizens have added their voices to alert us to the potential hazards of gene technology.

It is only mildly reassuring to know that in every country where genetic research now takes place, the government, in tandem with various private agencies, supervises its application. At present, genetic engineering is still in its infancy and these agencies can do their overseeing successfully. As time goes on, however, much more than the current supervision will be needed. Experience has taught us that ethics and morality cannot be enforced. These qualities must be voluntarily adopted by the people who are involved or they will be circumvented. In the case of genetic engineering, we must think far into the future that we are helping to create today.

Genetic engineering of brain chemicals can obtain similar results as those created by street drugs. The United States spends approximately $75 billion annually to control the still thriving street drug trade. The situation in future genetic engineering may not be different from the difficulty drug enforcement agencies have encountered in their largely futile attempt to eradicate the street drug trade. Looking further, we find that the United Nations has had only limited success in preventing the international trade in the materials utilized in making atom bombs. Current events indicate that we tend to overestimate our capacity to control our future. Is it reasonable to question if past misuses of science may be forerunners of what could happen to gene technology? At present, bioethicists still have considerable influence. However, what will they be able to accomplish later, when gene technology becomes a multimillion dollar industry, supported by extensive advertising and subject to political pressure? We need not go further for the answer than to look at the tobacco industry today.

In less than fifty years, geneticists will be able to manipulate our entire human genetic make-up. Entrepreneurs, not primarily concerned with human well-being, will then be in a position to control a variety of industries which will have evolved from genetic research. For this reason, I urge that we pause to ask, "Doesn't it make sense to say no to further genetic engineering research until we have gained the social maturity necessary to insure the safety of our species?"

Let us turn to our daily local, national, and international news and seriously ask ourselves, are we, at the present time, qualified to play the near-divine role that redesigning life would give us? Before we continue engineering the genome, we must, in the words of a modern analyst of our American society Hedrick Smith, "develop new ways of thinking about ourselves . . . what is needed above all is a new mind-set." The key words here are "a new mind-set." Individuals ranging from Pope John Paul II to writers of letters to the editors of daily newspapers call for something akin to a new mind-set for the world's political and industrial leaders—and for you and me. Let us not confuse a new mind-set with the many superficial remedies to cure human ills that are prescribed almost daily. Since the beginning of written history, a plethora of suggestions have been proposed on how to best heal an ailing world. These failed because the reforms suggested did not include a new mind-set. A new mind-set, above all, requires us to examine the basics of our human nature. Since they are not always pleasant to face, it is comforting to gloss over them. We no longer can afford to do this, now that we are gaining the ability to alter human nature by means of gene technology. Searching for the components of a new mind-set is what part of this book is about.

Until we gain a new mind-set, there are better ways our molecular biologists can use their impressive talents. Infectious diseases are making a comeback world-wide. These diseases, once thought vanquished, have resurfaced and gained increased resistance to our vaccines and antibiotics. Ebola, HIV, hantavirus, and bacteria resistant to antibiotics remind us that microbes remain our deadliest foes. From 1989 to 1992, death from infectious diseases increased by an estimated 58 percent in the United States and the percentage is steadily increasing. Nobel laureate Joshua Lederberg of Rockefeller University in New York City believes, "the war [against infectious diseases] is winnable but nobody was on watch and we have become complacent."

I do not suggest that we discontinue the genetic research that has the potential to disarm drug resistant microörganisms. A recent report suggests that the human immune system may receive a life-saving boost by means of gene technology. Nor would I recommend that we halt already promising research with DNA vaccines that help our immune cells fight a wide array of viruses and bacteria, including HIV. Some cancers are caused by defective genes, and there is hope that we can control this scourge of humankind by gene repair or substitution. However, at this time, medical gene technology should be strictly limited to the prevention of disabling human diseases.

But even this may be risky. Today, without a new mind-set, even our trusted hospitals have ethical problems. A recent example reported by the news services described what it termed "a massive fraud in medicare . . . Sutter Community Hospitals of Sacramento, California, agreed to repay the government $1.265 million to settle allegations that it had billed Medicare illegally for using experimental medical devices not covered [by Medicare]." The report continues to state that 125 other hospitals nationwide also were under investigation "for making false claims." I hesitate to use this example, since it should not reflect unfairly on the medical profession which, on the whole, has a good record of ethical commitment. Instead, these violations are symptoms of our contemporary amoral culture. Many people will agree that gene technology may get out of hand within such a culture. They believe that the abuses we see in illegal drugs and nuclear arms technology may foreshadow illegal uses of genetic engineering in the future.

We can learn what may happen to gene technology from the increasing number and complexity of fraud cases in high-tech science. In one medium-sized city in Southern California, three people stole phone numbers from over 650 cellular phones recently. They sold them to students, gardeners, business owners, and others for $100 to $200, allowing the callers to make thousand of dollars worth of phone calls all over the world. The thieves accomplished this by a process the police call "cloning," a word that also applies to copying life-forms in molecular biology. Thieves, some standing on freeway overpasses, use special electronic equipment to record the serial and identification numbers that are transmitted over the airwaves as callers talk on their cellular phones. Let us stop to think what would happen if future thieves used such sophisticated methods to commit crimes in the field of gene technology. Can we exclude the possibility that illicit genetic redesigning mills might spring up all over the world?

We have little basis to doubt that, given our way of living and present values, some people will exploit gene technology. Humans are a gullible species. Let us take the example of the recent popularity of body shaping machines. About $150 million worth of these have been sold in less than a year. George Salem, associate professor of physical education at California State University at Long Beach concludes, "using such (machines) in the manner prescribed by the manufacturer may, in fact, do more harm than good." One news commentator wrote, "people are looking for magic, not reality, and that is what the information market preys on." We are not apt to change in the future. Fashion models set the pace. In time, the magic may involve genes instead of washboard stomachs, firmed-up muscles, and weight reduction.

The history of science suggests that once we start the exploration of a new technology we tend to take it to wherever it will lead us. Rocketry took us to the moon. Genetic engineering may take us to attempt to redesign ourselves. Many people are horrified at that possibility and doubt that geneticists would ever be allowed to redesign human beings beyond making medical corrections. They are wrong. There was a time when people were horrified at the thought of using poison gas and germ warfare. Although international agreements have banned germ and gas warfare, some nations are still stockpiling these instruments of destruction. Today, some people would find it difficult to believe that not too long ago involuntary sterilization was mandated in the United States. We shall have more to say about this later.

Genetic engineering lends itself for use in both peace and war. If we permit it to take its natural course, which is to develop it to the outer limits of its potentiality, we may, in time, succumb to a mind-altering gene technology probably in the name of greater efficiency. People who are genetically engineered to cope effectively with life's problems will lose their ability to overcome obstacles. Throughout our human history, overcoming obstacles contributed to the enlargement of our human brain. Our brain enabled us to perform a miracle in formulating such deceptively simple question as, who are we and where are we going? No living creatures other than humans can ask such questions, let alone try to answer them. A genetically engineered mind will not require the characteristics that, over time, have made us human. With disuse, the qualities that made us humans will cease to exist.

Because my words may seem unnecessarily alarming—or even scaremongering—I would like to add some additional considerations. Let us not make the mistake of thinking that governments or bioethic groups can prevent genetic redesigning from becoming a weapon of war. We should be aware that potential abuse of gene technology could be as dangerous as splitting the atom that created the atomic bomb. Nuclear fission led to the scenario of Hiroshima and Nagasaki in World

War II, in which over 200,000 people were killed. Two more bombs had been built during World War II and the threat of using them helped convince the Japanese to surrender. However, imagine the conflagration that might have occurred had Japan at that time also possessed atom bombs with which to retaliate. Did Enrico Fermi and those who awarded him the Nobel Prize have any notion where his experiments with radioactivity might lead? Raising the possibility of genetically redesigning prisoners of war, or whole captured populations, is not scaremongering. Had anyone suggested that the initial experiments with nuclear fission would lead to a future devastation on the scale of Hiroshima and Nagasaki, the person expressing such a view would have been called an alarmist.

We do not need to bring war into the picture. One of the powerful lures of genetic engineering is found in human nature itself. Better than anything else, gene technology may serve our human fondness to change ourselves physically and mentally. We have not resisted doing so since ancient times. Every one has fun at a masquerade party. In the past, some humans have gone to extremes to gain new or enhanced personality traits. Throughout human history people have appealed to God, gods, or the spirit world to help them change themselves. They have worn devil masks, dressed themselves in animal skins to adopt the characteristics of these animals, and pierced noses and ears for adornment. Religions world wide reflect the human wish to be different or, as it is sometimes put, to be "born again." Christians find comfort in the idea of a different existence in the hereafter, while Eastern religions eagerly look forward to a series of altered lives until they reach Nirvana. To get into Heaven or reach Nirvana takes effort, avoidance of temptations and self-discipline. Genetically redesigning ourselves could some day make the task effortless. Then we, ourselves, can do what, so often, we have requested supernatural higher powers to do for us. Genetically redesigned people would have no need for religion, nor would they tolerate it if they, themselves, were able to "play God."

We have heard the phrase, "going from the sublime to the ridiculous." This we do as we now proceed from praying to examining the purpose of cannibalism—an activity repulsive to us. It has been practiced in such divergent places as Africa, South America and the South Pacific islands. The objective of eating another human being was not often prompted by hunger, but in order to transfer personality characteristics from one person to another. A cannibal's goal was to create personality changes for those warriors who consumed their enemy. A person's organs were thought to possess attributes such as strength, courage, and cunning. By consuming specific organs, cannibals believed they could transfer their victims' desirable qualities into their own bodies and minds. In ancient Greece, cult members who participated in some of the Dionysian rites consumed human flesh in an attempt to attain a semblance of godliness. Human motivation has not changed very much over the millennia. Our modern "cannibalism" has merely moved from the physical to the social and the psychological. If we recognize some truth in this, can we deny that it would be safer to wait until we have developed a new mind-set that excludes all types of cannibalism before we go much further with the application of genetic engineering?

CHAPTER TWO

OUR REPLACEMENT BY
HOMO GENETICUS

It is not an easy task to make a case for saying no to further research in genetic engineering in a world enthused with the idea that we are discovering the basic secrets of life. In this chapter I shall try to complete my arguments for saying no and to share with the reader my belief that to say yes to genetic engineering prematurely may lead to the extinction of our present species. A new, genetically-engineered species, *Homo geneticus*, is waiting around the corner to take our place.

No battles will be fought. No Darwinian natural selection laws will be violated. The take-over will be caused by our own human nature. As I have tried to show in the preceding chapter, our Achilles heel is that we are easily seduced by the idea of changing ourselves and, in the past, have even resorted to cannibalism to add new qualities to our personality make-ups. Genetically engineering ourselves may not appear any more hideous in the future than cannibalism did to its practitioners in the past. Genetically engineering ourselves has a deep-seated inherent appeal to us that most of us would prefer to deny.

As far as I know, this point has not been brought out clearly before in connection with genetic engineering. But actually, genetic technology's appeal should not surprise us. In the previous chapter, I mentioned people's eagerness to create their own personality changes by means of alcohol, drugs, and medications. I now wish to add the powerful role of vanity to the equally widespread wish people have to escape from themselves. Indeed, vanity may play the largest role of all. Gene technology offers us the golden key to the fulfillment of an ancient dream—to satisfy our vanity by looking different from the way we were created. The notion to change our physical appearance began with the loin cloth or the biblical fig leaf. Initially, dressing may have involved only adornments with leaves and flowers. Later we covered our nakedness for a variety of other reasons. One, of course, was for warmth. Another was for our vanity—to dress up in finery of feathers or silk, depending on our culture. In modern times, the same vanity prompts some ladies to do the opposite when they choose to wear only bikinis. Quirky humans that we are, we dress for vanity and undress for vanity.

Certainly, vanity will give genetic engineering a powerful impetus. Let us give vanity a few more thoughts, together with illustrations, because vanity may become a driving force in leading to genetically engineering humans. It takes examples from the past to help us recognize this. Pre-Columbian Meso-Americans artificially elongated the shape of their skulls to conform to the concepts of the fashions of their time. The Maori of New Zealand decorated their entire bodies with elaborate tattoos. Up until the early 20th century, aristocratic Chinese bound the feet of female children so that as adults they would be unable to walk and have to be carried by servants, adding to their husband's prestige. The Bonyankole people in East Africa fatten young women to make them more desirable for marriage. The reverse happens in our Western culture. Here many people work hard to slim themselves down to make themselves attractive. Today, almost everyone wears some form of jewelry. Some people who can afford to do so pay plastic surgeons to alter their appearance. When the time arrives that genetic engineering can more permanently create these changes, social pressure to use this easier method will build up immensely.

What will impel us in the future to resort to genetically engineering ourselves? Think of how much time and money we spend on make-up, hair styling, shaving, and polishing fingernails or artificially elongating them. A multi-billion dollar cosmetic industry exists only because humans insist on changing their natural appearance. We can imagine that, when it is possible, specialists in cosmetic genetics will be in great demand, even for embryos still *in utero*. The amount of wealth that could be gained by genetic altering cosmetology is staggering. Accompanying this bonanza would be a great new opportunity for fraud.

Today, I read an article in our newspaper that had an unusual twist. It seemed to me that some day it might apply to genetically engineered cosmetology. The article's headline was, "Doctor to suspend organ enlargement." This was followed by: "One of the nation's busiest practitioners of penis enlargement has agreed to stop doing the operations pending a hearing on whether he may keep his license." The paper continued to report that the doctor whose enlargement procedures had been widely advertised was "an immediate danger to the public's health, safety, and welfare." Imagine the much greater danger if some errant future doctors used genetic alteration to create irreversible physical changes that endangered public health. We already have heard much about implanted silicon breasts and surgical breast reshaping. What if we could play God by genetically engineering people to have these cosmetic alterations? But these alterations hardly compare to the lure of extending life. An "immortality gene" has recently been discovered that could greatly lengthen human life-span. In view of the huge amount of money some people would be willing to pay for a longer life, life-prolonging genes would certainly be immensely popular, even if they were illegal. Let's not even think of the overpopulation that might ensue!

If gene technology can prolong life, it can also shorten it and this might be found convenient for unscrupulous persons in whose hands genetic engineering decision-making rested. The fact that the necessary techniques are yet to be perfected is not the point here. We are looking at the Big Picture, where life is not measured in decades. We are building the future now by what we permit genetic researchers to do in our lifetime. We started the ball of genetic engineering rolling. We are on the way. One step will follow another. No doubt that within the next hundred years easier and more efficient ways will have been found to genetically manipulate human traits and emotions. That is when a new species, *Homo geneticus*, branching out from us, is likely to replace us.

A legitimate question at this point is: "Are inherited tendencies really as influential as some people now claim? Can we not override them by shaping a child's personality with environmental interventions?" Research studies have already linked a large variety of personality traits to an individual's genetic make-up. Such a close relationship between heredity and behavior would be hotly disputed by nature/nurture sociologists were the evidence less compelling. Already we know that genetics plays a role in determining whether an infant will develop into a thinker or a doer, be outgoing and active, shy and withdrawn and, of course, short or tall. Before birth, we can use biological techniques to learn whether an embryo will become a boy or a girl. Some day, no doubt in the lives of today's children, we will have the ability to see and, if we desire, alter the entire blueprint of a an infant's genetic predisposition—physical, mental and, emotional—before it leaves its mother's womb.

Nevertheless, it is right to question the absolute power that some people attribute to heredity. Reliable studies confirm what sociologists have claimed all along. Genetics, alone, does not determine an individual's personality traits. The mix of genetic traits and environmental exposure combine to make up human behavior. Some genetic traits are actually elicited by specific environmental conditions while others are suppressed by them. Heredity and environment have feed-back loops. One informs the other what to do. That is why we can afford to say no to genetic engineering for now and wait until we can more successfully use environmental ways to improve our society.

One is apt to ask at this point, why, in spite of all our indoctrination, education, philosophy, and psychology have we so far failed to create better kinds of people? One answer is that we are unable to agree on what "better" means. Even inhabitants of the smallest village, let alone the world, fail to agree on what traits "better people" should have. Religions also differ according to their own interpretations of their scriptures. The differences of opinion can only become more divisive when genetic engineering plays a larger role shaping people's behavior. Imagine redesigning people to avoid culture shock as they move from one culture to another or when they change religions!

It is clear that we must gain a new mind-set by means of available environmental techniques. Let me present one of many examples of environment's success. It offers us a valuable insight because the work was done with a population known as difficult to influence. Craig Ramsey, a University of Alabama scientist, reported at a 1996 meeting of the American Association for the Advancement of Science that a group of high risk mentally retarded children faced a lifetime of difficulty in learning. They gained 15 to 30 IQ points by participating in intensive, early intervention programs. "In highly vulnerable groups, we cut the risk of mental retardation by over 50 percent," Ramsey stated. Another study showed a strong correlation between a person's ability to cope mentally in advanced age and their environment. Older persons exposed to new facts who continue learning tend to maintain mental acuity and capability far beyond a matched group who did not have the environmental opportunity or stimulation. Impatient advocates of genetic engineering might overlook certain facts, such as the finding—recently confirmed statistically—that excess concentrations of lead in their bones causes aggressiveness and delinquent behavior in children.

The investigators concluded that environmental lead exposure—a preventable occurrence—should be included when considering factors contributing to delinquency.

In spite of these facts, we must be aware that the lure of gene technology shaping remains always behind our backs. We are reminded of identical twins studies that have demonstrated heredity's close tie to behavior. Gene technology would, in time, offer a huge, money-saving short-cut for rehabilitating some individuals indelicately called "socially unfit." These are the recidivists who play a significant and repetitive role in disrupting society. It is easy to see that life in prison and capital punishment have a key factor in common with gene technology. In each case these persons are prevented from adding their genetic characteristics to the human gene pool. Considering this, the distinction between capital punishment, life imprisonment, and genetically redesigning the human race is really not as great as one might think.

Merely altering small amounts of dopamine, serotonin, norepinephrine, and acetylcholine in the brain makes an enormous differences in the way a person would feel. Only a slight change in the quantity of the brain's neurotransmitters would guarantee that a person has a lifetime of happiness. Creating such changes by means of genetic engineering will be possible in the foreseeable future. Techniques that greatly simplify how to accomplish this will be discovered. Some day it may be as easy as infecting a person with harmless microorganisms, into which alternate human genes have been inserted. Even today, we can use this method to create transgenic animals.

To a large extent, the genetic blueprint for all forms of life is the same and that contributes to its potential for abuse. For example, 8000 genes have already been found in one species of worm and 40% of these have similarities to those of animals. In a few cases, human and worm genes are so closely related that human genes are able to restore normal functioning in worm genes that have become defective by mutating.

Those who are skeptical of what genetic engineering could lead to should turn to their encyclopedia and look up the word "eugenics." Eugenics is a late 19th century theory whose advocates claim that we should enhance the human race through selective breeding. It is in fact nothing less than a roundabout method of human genetic engineering. Eugenics was founded by Sir Francis Galton, a cousin of Charles Darwin, who noted that exceptional ability ran in some families and proposed the idea of human selective breeding in 1869. He also advocated sterilization of the "genetically unfit" without giving those so identified any voice in the decision—a practice made legal in Great Britain and the United States. Given the chance, those who wish to accomplish the same goal may use genetics as a replacement for selective breeding and sterilization.

We have covered some pros and cons regarding the use of gene technology to alter our species. Some well-meaning people in their enthusiasm for innovation might insist that since we are capable of playing God, let's do it to improve the human race. They may not realize that geneticists (or those who employ them) will always retain "ownership" of the altered personality they create. The geneticist's property rights will extend to the personalities of the client's descendants. Their genetically changed personalities might even be legally patented by the geneticists or their employers. Eventually, more and more people might serve merely as housing for various geneticists' works. Keep this in mind as you read the following seemingly improbable scenario.

Let's picture a group of genetically redesigned people—a future species of *Homo geneticus*—competing for their geneticists on a race track in the form of a sophisticated group intelligence test. Visualize the contestants cheered on like modern race horses by enthusiastic crowds of betting spectators. Pity the losers. They would have to be redesigned before being allowed to participate in future races.

What would these science fiction geneticists be testing for? The survival of the fittest, of course. It was the same way with the IBM chess-playing computer which Garry Kasparov, the world's leading chess master, defeated. The computer was reengineered each successive year it lost to a human chess master. It's chess rating went up steadily. At the time of this writing the computer can analyze 80 million to 120 million positions per second. It won one game, but lost the series to a human being, who could use the imagination the computer lacked. Who would have guessed a few decades ago that people could build a computer that can explore over 100 million chess moves a second? We are short-sighted if we do not look ahead at the unexpected and unbelievable. We should learn the probable fate of genetic technology from the French novelist and inventor of modern science fiction, Jules Verne (1828–1905). In his time, his novels about submarines, a trip to the moon, and circumventing the earth in 80 days (we can do it now 18 hours) were ridiculed as impossible.

If the lure of genetic engineering becomes irresistible, it is because it has the attributes of a fascinating intellectual game that tops chess. Humans may well then become the pawns in the game. As in chess, other pieces (probably early *Homo geneticus*) would represent the kings and queens of their genetic domains. To carry the analogy further, the bishops might assume the additional duties as politicians, and the knights, which in chess sometimes appear unexpectedly from a corner of the board, might be lobbyists. Alternately, the genetic redesigning game may take the form of bridge, poker, or Bingo. Many thousands of dollars worth of damage occurred when so-called "viruses" that disabled computers world-wide were playfully inserted into software by pranksters. Soldiers engage in field exercises they call war games. In the hands of dictators, real wars are sometimes waged to entertain the nation and keep would-be dissenters occupied. Future dictators would find redesigning human beings a rewarding game that gives them the advantage of boasting of their humanity and compassion. Instead of killing their prisoners of war or dissenters, as in the 20th century, they would only genetically redesign them. They would surely try to obtain good public relations by comparing redesigning people genetically in contrast to genocide.

Human nature was well understood by Buddha, who lived in the 6th century B.C. Now I will tell you a parable Buddha said.

> "Once there lived a wealthy man whose house caught on fire. The man was away from home and when he came back he found that his children were so absorbed in play, they had not noticed the fire and were still inside the house. The father screamed, 'Get out children! Come out of the house! Hurry!' But the children did not heed him.
> "The anxious father shouted again. 'Children, I have some wonderful toys here; come out of the house and get them!' Heeding his cry this time the children ran out of the burning house."

With the power of genetic engineering in the hands of unscrupulous people, our world will be a burning house. Most people, fascinated by the possibilities of genetic engineering, are oblivious to the potential blaze. I am trying to warn us before we are driven to extinction by *Homo geneticus.*

Within the next century, the entire genome will be precisely mapped, enabling geneticists to move our genes around on the DNA chessboard we carry in every one of our cells. At that time, the human brain will become the most fascinating toy on earth. It is possible that playing games with genes would fall under the heading of "brain enhancement" or even "clever adaptation" to make the idea acceptable to the public. Then the new creature with a genetically redesigned brain, the *Homo geneticus*, will be on hand to replace us. *Homo sapiens* would become an endangered species. *Homo geneticus*, with his genetically engineered compassion, would feel duty-bound to assist our present pitiful species of humans to become extinct. A few old-fashioned humans like us may be retained for comparison to demonstrate *Homo geneticus'* achievements. We do the same with our museum dioramas of Neanderthals.

If genetic technology gets into the wrong hands, Buddha would agree, "our house would be burning." We, who are now living, must find a way to save ourselves. We are vastly better for the world than a redesigned automated species we have called *Homo geneticus*. We can survive if we stop playing games, take our eyes off our genetic toys, and seriously think about our future.

But let us look at it another way. Let's look at world-wide malnutrition. Malnutrition caused by caloric deficiency afflicts an estimated 400 million to 1.5 billion of the world's people. Even in the United States, *National Geographic Magazine* has reported, undernourishment exists among an estimated twenty million citizens. In Columbia, South America, nutrition-related diseases claim two of every five children. In Bangladesh, the specter of starvation stalks the land. In some African countries similar conditions exist. In the war against starvation, cross-breeding of food crops continues as it has for many years. To increase the world's protein, livestock breeders have transplanted embryos from prize cows to substitute mothers. A single cell protein can be grown in a mixture containing a petroleum derivative. Some nations are harvesting the oceans' abundant plankton crustaceans—krill—to convert into human and animal food.

Unfortunately, this is not enough. Starvation and malnutrition continue. Some people point out that our only hope lies in the use of gene technology. Transgenic plants and animals created by genetic engineering offer the best chances of success in our fight to provide new varieties of pest-resistant foods, cure or arrest some of our worst and deadliest diseases and finally win our ancient battle against harmful microorganisms. If our house is burning, shouldn't we try to extinguish the fire? Morally, are we compelled to risk increased research in genetic engineering to help put out the fires of hunger and starvation?

Few would deny it if somehow they could be assured that this level of genetic research would not, in the end, lead to our capitulation to *Homo geneticus*. But how can we be sure of this? I have argued that it is very doubtful that we can prevent genetic engineering from progressing eventually to attempts to alter humans—or some humans. I have used two chapters to argue that even if we intend to limit research to prevent inherited diseases, genetic technology is likely to drift on—under its own momentum—to an unpredictable future that may involve redesigning human personalities. Beclouded by all the pessimism, there is now also a new reason for optimism. For the first time in our history, we have a powerful incentive to gain a new mind-set that has existed thousand of years in our prayers but not in our lives. Can we rally humankind under the banner: "Make our world safe for gene technology?" This will be our concern in the chapters that follow.

CHAPTER THREE

WHAT'S GENETIC ENGINEERING ALL ABOUT?

Let us now examine some of the intriguing facts about the composition of the human genome. Within the nucleus of every one of the approximately ten trillion cells in our body, we carry about 100,000 genes tightly compacted on two winding strands, known as the double helix, composed of deoxyribonucleic acid. Fortunately, we don't have to worry about this term because it has been abbreviated to DNA. The genes on the human DNA are located in 46 bundles called chromosomes. Bacteria have only three chromosomes, but this gives us no reason to feel superior—turkeys have 82. By 1940, scientists had established that the DNA molecule is composed of chemical building blocks in rows of repeated sequences called nucleotides.

Humans have four billion pairs of nucleotides. We often read about a billion when it comes to questions of financing or costs in dollars. Therefore, we may forget that a billion is a very large sum. To grasp the meaning of a billion, let's remember that it is the product of one thousand times one thousand times one thousand. Four times that amount of nucleotides are compressed into each DNA molecule. The invisible DNA molecule would extend close to an incredible five feet if it were stretched out to its full length. Our nucleotides were assigned the letters A, C, D, and G after the first letter of the chemical names of their constituent proteins. When they were first identified, it seemed most of these proteins had no function and just took up space. They were thought of as meaningless filler on the DNA. Later it was found what had been thought of as "DNA junk" by molecular biologists had a discernible structure that somehow resembled a language. Now that some of its patterns are gradually being discovered, the seemingly surplus DNA areas retain a sense of mystery. The chances are that in time we shall learn that these nucleotides play a role that we as yet do not recognize.

Scientists had a similar experience in exploring chaos which at one time was viewed as pure randomness. On a closer look, computer models were evolved that replaced the idea that chaos was mere meaningless jumble. We tend to call things meaningless if we find them useless to us.

One of the important discoveries of gene function was the existence of suppressor genes. These genes have been called "guardian angels" by molecular biologists. They cause injured cells to self-destruct before they cause cancer or other bodily malfunctions. Some genes play the roles of master switches. They send chemical messages to other genes by way of enzymes that cause physical or mental changes to take place at various times in the life of an organism. Among these are the genes that turn on the secondary sexual characteristics by triggering the flow of hormones that cause bodily changes during adolescence. Genes time many bodily functions and account for the observation that we have within us a number of physiological clocks, including one that ticks away the number of years we live.

The structure of the DNA was first described in the journal *Nature*, April 25, 1953, by Francis H. C. Crick and James Watson. In 1962, they shared a Nobel Prize in Medicine for their discovery. Both have written books describing their experiences in the pursuit of the secrets of inherited characteristics. Watson's best selling book *The Double Helix* (1986) describes how the double strands in the DNA molecule contain the structure of the human DNA. A book by Crick, *What Mad Pursuit* (1988) looks back on events in his research.

The human genome, a term used to describe the sequence of all the genes on the DNA molecule, carries the exact blueprint of the genetic make-up of our species. A person's general appearance, such as color of hair and eyes, height, as well as genetically determined aspects of health, abilities, and personality, lie within the DNA. Biologically, minor variations in the DNA code bestow on us the traits that make each person unique. We also carry within ourselves genes that have no affect on us. We transmit them to our descendants, so that a characteristic that did not show up in our lifetime may shape theirs.

This happens because organisms have both dominant and recessive genes. The dominant genes give us our own characteristics. But if two people mate who both have the same recessive gene, the characteristic that gene expresses may show up in their children. A single defective gene may sometimes be responsible for a specific genetic illness, such as the gene that causes Huntington's disease. This is a progressive hereditary disease that usually has its first symptoms in middle age. It is devastating. Patients ultimately lose their ability to care for themselves both physically and mentally. Other illnesses and behavioral traits influenced by genes are usually polygenic—that is, affected by a combination of a number of genes, each playing a specific role. Some genes perform their tasks in unexpected and complex ways. As I have stated earlier in this book, a person's environment, such as exposure to pollutants, emotional stress or poor health habits may activate a genetic tendency. We must view environment and heredity as partners.

So far, I have said no to genetic engineering that involves personality changes because much that we could do to improve our health and well-being can be accomplished by environmental factors that we fail to exploit. Many molecular biologists think that we may never be able to manipulate the genome to produce an Einstein or a Beethoven. Genius, reverence, awe, and other higher level human qualities and feelings may, for an indefinite amount of time, remain beyond the geneticist's reach. Nevertheless, geneticists continue to map the uncharted genome and they may eventually accomplish what is presently inconceivable. Scientists estimate that within the next thirty to fifty years molecular biologists will complete the huge task of locating the exact position of every gene in the human genome. Once our genome has been mapped, human genes can be catalogued and accessed in much the same way as books in a well-stocked library. Geneticists will then be able to borrow, exchange and replace genes using the same proteins, called restriction enzymes, that serve bacteria as chemical scissors.

The bacteria use these proteins to cut out of themselves those of their own genes that are susceptible to antibiotics. Mutated genes unaffected by the antibiotics then replace the susceptible ones. To our consternation, these bacteria can become immune to the antibiotic biological weapons we use against them by this process. Unfortunately for us, they pass their immunity on to following generations and, worse, also to different species of bacteria. It is really astonishing that molecular biologists can extract the restriction enzymes from bacteria and use them to cut out genes from the DNA of any living thing—any animal, plant, fungi, or human. The basic mechanism of nature's grand scheme used to pass on genetic characteristics from generation to generation is the same for all forms of life. This brings to mind that, in a sense, all of life belongs to one huge family that would be destroyed if some members of this family were not sacrificed in order to keep other members of a different species alive.

In order to appreciate what is involved in gaining a new mind-set, we must be aware of the roles of both genetics and environment. Beyond the requirements of physical healing, some genetic engineering can reach deeply into the realm of human behavior. A 1993 study in *Science* of more than 1,500 pairs of twins suggests that divorce has now joined the growing list of human behaviors that have been found to be, in part, influenced by genes. Researchers have identified a variety of inheritable traits that relate to an individual's capacity for happiness, job stability, neuroticism, impulsiveness, and sexual behavior. Danish researchers have added even such unexpected traits as persistent bed-wetting to the growing list of behaviors in which genes may be implicated. Nevertheless, environmental influences often play a more critical role. That is why research of personality traits from genes should be accompanied by corresponding research in the role played by environmental factors. It is important to know whether or not changed environments alone could produce desired results. This research must have priority over genetic investigation since environmental change is our only hope of ever achieving a new mind-set. The only alternative left is a new "gene-set."

As a psychologist, I know that a large number of people have found happiness by turning over a new leaf, changing life styles, or using nonviolent outlets for built-up irritations. As I shall show later, a new mind-set can be achieved by the strengths and inner resources that we already have but do not draw upon. It is important to remember this as we face one of the most troublesome and persistent problems of our century—the increasing violence of our population.

The problem of ascertaining the causes of violence is a touchy one. A recent issue of the *Psychological Science Agenda*, carried the headline, "Violence Research Produces Controversy." The article stated that the U.S. Public Health Service canceled funding for a conference on genetic factors in crime because of protests by civil rights leaders. They feared that the results of the research might be misused to increase bias against minorities. Recently emerging disagreements indicate what is ahead. Charles C. Mann writing in *Science* (1994) predicts: "There are few certainties in life. The uproar surrounding attempts to find biological causes for human problems will continue." Under the heading, *War of Words Continues in Violence Research*, Mann reported on the acrimonious controversies over nature's versus nurture's role at the 1994 annual meeting of the American Association for the Advancement of Science.

These are mere beginnings of the dissentions that will occur if an influential scientist or politician proposes that we attempt to use genetics on a large scale to make people content or easier to live with. It reminds one of Aldous Huxley's *Brave New World*, first published in 1932. In Huxley's novel, the Director of the fictitious London Hatchery and Conditioning Center found a method of producing a multitude of identical twins in order to create social stability. Toward the end of his book, Huxley describes them as shouting in unison, "parrot-fashion . . . intoxicated by the noise, the unanimity, the sense of rhythmical atonement." It is clear that Huxley visualized *Homo geneticus*. The future will decide if Huxley was merely an entertaining author of science fiction or a true prophet of what will come about in the "Age of the Genome." From what I have said so far, it appears that Huxley may have been another fiction writer whose fiction could some later day become fact. We have reached the stage now where we must take responsibility for the future direction humankind will travel. Without achieving a new mind-set, it will be Huxley's.

We need not even look at the future, since we now already confront problems raised by genetic discoveries. One example of this is found in concerns about genetic discrimination in employment. How will employers and insurance companies deal with individuals whose genetic profile indicates the potential for developing mental or physical diseases? The media already discusses concerns over genetic labelling and pictures a potential genetic underclass. With reason, people fear that turmoil, bitter political and religious confrontations and, perhaps, even physical conflict will take place over who has the right to manipulate the human genome. The misuse of our emerging god-like power to redesign life will become the dominant topic of controversy in the next century. We can also anticipate competition and dissension in regard to the various methods for manipulating the genetic nature of all life forms. For example, the method of cloning organisms may some day compete with recombinant gene technology in an aggressive struggle to gain preferences and privileges. Controversies will arise regarding whether a given technique is safe to apply or whether it should never be applied. We cannot remain apathetic.

In the previous chapters, I suggested that this is not a time for gullibility. I mentioned that present regulating agencies give us a false sense of security as far as gene technology is concerned. In our country, the National Institute of Health (NIH) and the U.S. Food and Drug Administration (FDA) regulate gene therapy. The U.S. Department of Health and Human Services oversees both agencies to insure safety in the application of human genetics. Other groups also hope to persuade governments to enact laws prohibiting the unrestricted experimentation and manipulation of human genes. During the first quarter of the 20th century, when prohibition outlawed all intoxicating liquor, everyone who wanted a drink knew where to go to get one. In the end, a criminal element providing alcoholic drinks illegally to the public gained power and influence. Today it is the illegal street drug trade that follows this precedent. No one can rule out the possibility that in coming generations it may be an illegal, clandestine genetic engineering industry that provides demanded services.

We are entering an age where, in theory at least, we shall be able to genetically redesign ourselves. In the previous chapters we discussed reasons why the temptation to do so will be difficult to resist. How can we obtain a new world outlook that differs enough from our present one to make it less likely that genetic engineering will be exploited for greed and selfish ends?

To answer that question, we must trace our footsteps back and discover how we arrived at our contemporary dissatisfactions, hates, and violence that have taken on the proportions of a revolution against our very selves. The roots of this revolution can be found in our earlier Western history. Events at that time created repercussions that are still felt across the modern world. The following chapter explores the consequences of events that first were hailed as major advances in civilization. Understanding them is necessary in order to discern the impact of gene technology in the years ahead.

CHAPTER FOUR

OUR CONTEMPORARY REVOLUTION AGAINST OURSELVES

Along with the can-do optimism and accomplishments that accompany our technological breakthroughs, our Western civilization must also cope with disturbing and seemingly insoluble problems in our daily lives. In this chapter, we shall search for the reason that opposition to the influences of Western civilization surfaces in many parts of the world. People vainly try to find a meaning in life beyond success in their employment and professions. Reasonable decisions on how to apply genetic engineering to human needs are difficult to make in a world of dissension and disillusionment. The present incongruous mixture of optimism and discontent cannot avoid causing bitter conflicts over the use of gene technology in the Age of the Genome.

Today, there is hardly a newspaper editorial page that fails to tell us what is wrong with our society. Religious leaders of many faiths call our attention to what they identify as a contemporary spiritual and moral decline. Sociologists point out that the mobility of our society deprives people of the traditional guidance and support of their immediate and extended families. Signs of our contemporary malaise include the rapid increase of violence, the widened generation gap, and the destructive drug culture.

Multiple factors contribute to our sociological problems. Usually they are complex and subject to a number of interpretations. Most people would agree that the flow of past events in the history of Western civilization has contributed to many of the problems we face today. Explanations for today's difficulties using a time frame of one or two generations do not permit us to view them within the larger framework of their chronological development. An awareness of their historical background will not solve them, but may provide the perspective we need to gain a deeper understanding of how they came about. This could lessen the danger of passing on distortions and oversimplifications to the generations who will live in the Age of the Genome.

Setting historical time frames for sociological events has to be arbitrary. Rarely in human history can we point to a specific event and say, "Here's where the problem began." Nevertheless, let us look back to the period in Western civilization known as the Middle Ages, which followed the collapse of the Roman Empire in the 4th and 5th centuries. This period was not as homogeneous as its other name—the Dark Ages—implies. There were times of great intellectual ferment and innovation in the arts and literature in the late Middle Ages.

The dominant system of social structure consisted of three classes—nobility, clergy, and serfs. Land was owned by kings and nobles and worked by serfs. Sometimes land was granted by overlords to vassals who swore allegiance to them in ceremonies that cemented their personal relationship. The feudal system was essentially a military organization with knights as warriors. Christian and military ideals created the morality of the times. Chivalry was an ethical ideal that required the knights to take vows of fealty to their lord and, in theory, protect the weak and poor. According to tradition, one of the objectives of knighthood was to impress women of their class with their exploits of skill and bravery. This was often accomplished on horseback in tournaments, where knights knocked other knights off their mounts by crashing into them with poised lances. Throughout history, the human can-do spirit finds outlets that seem strange to people living at a different time.

The intellectual force in Western Christendom that held medieval civilization together was called scholasticism. It consisted of theological teachings mixed with Greek philosophy. This was a time of the exploitation of agricultural workers, religious dogma, and superstition. There were ceaseless raids carried out by nobles who obtained their land and power largely in battles with one another. Nevertheless, feudalism offered the people, including agricultural workers, a personal identification that is missing in our contemporary culture. The paternalism of lords and clergy gave serfs a sense of connectedness within their own microcosm of society. Even as serfs were exploited, they found security in knowing who they were and what they should believe. As commerce increased in the late Middle Ages, people were drawn into towns and cities, where merchants gained political power and formed a fourth social class.

After merchants had gained increased social status, an intellectual revolution occurred in Europe in the 15th and 16th centuries. The revolution began in Italy, but became known by the French word for rebirth, "Renaissance." Independent city-states increased their political and economic power during the prevailing period of relative social stability. Urban centers of population established wide contacts with other cities and states. The revolution was hailed as a liberation for the people of Europe. It was a time of brilliant accomplishments in scholarship, literature, and the arts. It provided people with a new world-view of human destiny. The intellectuals living at the time of the Renaissance would not have believed that anything but good could result from their new sense of freedom. Ironically, one of the most benevolent periods in European history, noted for the liberation of the human spirit, may have helped shape the distinctive characteristics of our contemporary social problems.

In 18th century Europe, the Age of Reason followed the Renaissance. Intellectuals attacked religious intolerance and censorship. After hundreds of years of hearing themselves referred to as sinners, people were told that they themselves had the capability to make the right decisions about their lives. A contagious optimism swept through Europe during the years known as the Enlightenment. People believed that if scientists had sufficient information, their rational minds could solve all human problems.

Innovative ideas made it an exciting time for intellectuals. The people of the Enlightenment no longer believed that they lived on an earth around which all planets, stars, and galaxies revolved. The English mathematician Sir Isaac Newton, born in the same year that Galileo died, discovered the laws of universal gravitation. His many accomplishments included the formulation of his three laws of motion, which helped make physicists aware that order exists in the universe. When evidence and proof replaced intuition and false certainty, science flourished.

The new doubt about the nature of reality elated some people, but disturbed others. The philosopher Descartes lived at about the same time as Galileo (1564–1642). Descartes believed that nothing should be taken for granted and that even human existence must be doubted. Later, the influential Irish philosopher and clergyman Bishop Berkeley (1685–1753) proposed that there was no existence beyond perception. He wrote: *"Esse est percipi"* (to be is to be perceived). He maintained that perception was the only source of reality and that only seeing justified believing. A tree isn't a tree until someone observes it. Some people were convinced that Bishop Berkeley was right. However, the idea that happenings become real only by virtue of someone's awareness of them breaks reality into fragments. In time, the fragments of reality that visibility represented took the place of total reality in many people's minds. This distortion of reality continued and still exists today. It helped to create the philosophical underpinnings of our contemporary problems.

The skepticism that accompanied the new freedom of thought caused many people to turn from faith to science for an explanation of life's meaning. In time, people decided that only doubt was capable of unearthing basic truths. The influential British philosopher David Hume (1711–1776) insisted that most of the principles of our knowledge had no justifiable basis. He believed that before any idea could be accepted, proof available to the human senses had to exist. Many people living in that period ignored—as some still do today—the fact that science, too, represents only a fragment of reality. The new doubt about the nature of reality elated some people, but disturbed others. In their confusion, people no longer knew their roles in life, as they had previously under feudalism. They felt compelled to search for evidence that their lives were tangible and genuine. They paid for their freedom with the loss of their only safe anchorage in an uncertain world—their unquestioning faith in God.

In 18th century England, the Industrial Revolution followed the Age of Reason. Life was exciting for inventors, entrepreneurs, and businessmen, but it displaced masses of people previously engaged in agriculture who left their rural homes to find jobs in the booming cities. Urban factories designed for specialization made the large-scale production of goods possible. An expanding population demanded more and better manufactured products and this emphasis led to an exploitation of the factory workers—men, women and, children. Many workers were forced to spend long hours in factories earning little pay for performing repetitive tasks on assembly lines. In time, the workers in the factories lost their human identities. Their bleak lives gave "invisibility" an ominous new meaning.

The conditions at the time of the Industrial Revolution created new social classes—owners, managers, and workers, with little responsibility towards each other. Exploitation of factory workers was accompanied by the dislocation of families crowded into tenements. Workers were considered little more than human robots. The person-to-person relationships that had held people together under feudalism no longer existed during the Industrial Revolution and the need for human interrelationships remains largely unmet today. In the United States, we pay the price for the lack of social connectedness by the proliferation of street gangs that attract many of our teenagers. These gangs offer their members identifying jargon, clothing, territories, strong internal loyalty, and even the togetherness provided by gang warfare. Such activities represent nothing less than a return to a kind of feudalism in miniature.

I request the reader's indulgence while I dramatize the lost feudal personal relationships by asking the following question: What would one think today of an employee who vowed lifelong allegiance and sank to his knees before the president of a business corporation who had promoted him to a middle management position? Under feudalism, this scenario was customary when a lord awarded a portion of land to a vassal. In the medieval promotion ceremonies, the overlord kissed the kneeling vassal and raised him to his feet. Thereupon, the vassal swore an oath of loyalty and vowed to be faithful forever. I am not recommending that we create similar promotion rituals today. However, our modern profit-oriented society is largely devoid of a sense of obligation and personal loyalty. An impersonal view of our fellow citizens is typical in business transactions, and in our civilization at large. In our milieu, we direct our warmer emotions primarily to issues. People, on the other hand, have become abstractions and symbols. This could lead to irresolvable controversies when the human genome is fully mapped and we are ready to apply genetic engineering.

The prevailing climate of anxiety and uncertainty of one's role in the world, precipitated by the need to obtain visibility, has led us to a 20th century movement called existentialism. It is a reaction to the previous view that the single fragment of reality—visibility—could, by itself, give life meaning. Both Christian and atheist writers contributed ideas to the movement. Existentialists denied that people required visibility to exist. They proposed, instead, that existence precedes essence, that is, first one becomes a person, then one becomes a particular kind of a person. Existentialism reversed Bishop Berkeley's "I see a tree therefore the tree exists" to "It is a tree and since I'm human it looks like a tree to me." Its theme consists of the idea that freedom and the lack of given rules create the need for us to be responsible for our actions. The controversial writer and philosopher Friedrich Wilhelm Nietzsche, who we shall discussed later, has been viewed by some as a forerunner of existentialism.

Existentialism gained considerable acceptance in Europe in post World War II France. Jean-Paul Sartre (1905–1980), French playwright and novelist, became a leader of the existential movement. In a popular novel, he described humans as lonely beings who were burdened rather than elevated by their freedom. He portrayed the current generation as drifting aimlessly through life. According to Sartre, peoples' anxiety could be viewed more accurately as anguish caused by the almost impossible task of giving meaning to a meaningless world. Existentialism offered an alternative to the notion that people must validate their lives by exhibiting their visibility. Instead, the new thinking suggested that people could give their lives meaning only by accepting responsibility for their own fate. Some people viewed existentialism as an elitist concept and existentialists failed to get their message across to the man in the street. People were not ready to accept the idea that they were responsible for their own problems but, instead, continued to rely on visibility to demonstrate that they were unique individuals. Today, responsibility still takes a second place to calling attention to oneself. We now experience a revival of "seeing is being."

Some scientists believe that the approach to consciousness is almost solely through vision. In his book, *The Astonishing Hypothesis: The Scientific Search for the Soul* (1994), Francis Crick, the 1953 Nobel laureate co-discoverer of the structure of DNA, argues that consciousness, "the essence of humanity" as he sees it, must rely on science to explain it. He recommends the use of visual neurobiology and psychophysics as approaches to reaching consciousness, because our neocortex has adapted vision as a primary resource for gathering information.

Consciousness, however, goes beyond information and awareness. An important aspect of the essence of being human lies outside the reach of physiological vision and that applies equally to *soul,* whatever meaning one may give it. Inner experiences rather than visual processing lead directly to the soul, while seeing does not always lead to inner experiences. The matter is complex, but if everything we could ever learn from science were absorbed by us, something would still be missing. Perhaps, in addition to whatever else occurs, this something plays a role in the meaning of life. J. J. Hopfield of the California Institute of Technology, in his review of Crick's book, writes, "The book is a heroic attempt to wrest consciousness from the minds of philosophers and place it in the hands of scientists." He referred to those scientists, I might add, who still believe in Bishop Berkeley's and Descartes' claim that physiology is the sole repository of reality.

Esse est precipi has invaded the halls of academia. Among college professors it is called "publish or perish." Excellent teachers who do not gain visibility through publishing professional books and articles find themselves in danger of losing their jobs. The use of violence, such as detonating bombs or hijacking planes, is a less academic way to gain visibility. However, it quickly creates visibility by capturing media headlines. Graffiti represents the pathetic signature of underdogs who yearn to become visible. Our contemporary violence is, in part, an expression of frustration that the utopias promised in the Age of Reason have failed to materialize. People are tired of waiting for the good life the Enlightenment promised and have fears that they have been misled. We have with us now what might be viewed as a revolution against ourselves.

In this revolution, the descendants of the European idealists who brought about the Age of Reason are now returning in ever larger numbers to the very authoritarianism, fundamentalism, and ultra-conservatism against which their 17th century forefathers revolted. They demand to know where they belong without having to demonstrate that they exist. Nikki R. Keddie, professor of history at the University of California at Los Angeles, makes the point in a postscript in *Contention* (1993), that resurgent fundamentalism is a modern trend which stems from widespread frustration with the failures of westernized nations to satisfy basic human needs. She has spiritual needs in mind that may be beyond what the Age of Reason was able to provide. Later, we shall try to identify what these needs consist of.

The Age of Reason encouraged democracy in governments and championed people's right to live in dignity, liberty, and freedom. But it failed to integrate visibility with important but less tangible aspects of life. The Industrial Revolution stimulated ingenuity and inventiveness. Nevertheless, it depersonalized and dehumanized many people living in urban Western Europe and the United States. It led to widespread unemployment during periods of economic stagnation caused by the replacement of manual labor with efficient, time-saving machinery. Can-do humans with nothing to do soon feel useless and become emotionally disturbed or embark on careers of crime. Some of our present dilemmas are rooted in the ground that the Age of Reason and the Industrial Revolution prepared for us. Notwithstanding the benefits we received from the Enlightenment and from urban industrialization, we have lost the feeling of human interconnectedness. The feudal system, with its unbridgeable class distinctions of nobles, clergy and serfs, still managed to maintain a togetherness of workers and aristocrats, all ruled over by an omniscient God. The interdependence of people created feelings of loyalty and obligation. In the modern world, loyalty, obligation, and a feeling of belonging fail to exist within our social fabric. Today, the best examples of a feeling of belonging can be found in ethnic, tribal, and racial chauvinism, and in the brotherhood of international terrorism. Something new must occur within our society if we are to avoid transmitting this chauvinism into the Age of the Genome.

Existentialism called attention to the rootlessness that accompanied freedom from obligation. The theme had an impact on a number of authors who wrote on mental health. The book *Existence* (1959), edited by psychoanalyst Rollo May, claimed that after people in Western civilization freed themselves from hunger and fatigue thanks to the use of agricultural machinery, they ran headlong into boredom and meaninglessness. The authors suggest that in order to transcend the existential vacuum—the emptiness in their lives—people must accept personal responsibility for their unsatisfactory contemporary condition and build a new, liveable society, as I have maintained throughout this book. Somehow, the chance to build such an improved world seemed to slip through our fingers. No doubt there will be people in the next century convinced that we should use genetic engineering and redesign ourselves as the only way to escape from our presently insoluble dilemmas.

The Age of Reason taught us that we have rational minds. Existentialists maintain that we are responsible for ourselves and our fate. We must try to integrate these ideas against a backdrop of indoctrination on political, health, financial, and religious issues given to us by aggressive people with their own personal biases who utilize the media to create visibility for themselves. Our own need for visibility increases our susceptibility to their persuasiveness. The malignant spirits feared by some indigenous people have reappeared in the guise of some media personalities. Television personalities become real people in our living rooms in front of our eyes. We may view them as modern knights of the media, unencumbered by any traces of feudal obligations. After the pictures on the screens fade, the TV personalities remain within us, haunting us to do their bidding. They use sensationalism in communicating that startles, shocks, and thrills while it misinforms. In spite of warnings against portrayal of violence and clamor for restraint, the violence which continues on the television screens represents a fragment of reality that becomes all of reality on the crime-ridden streets of our major cities. Mapping the human genome may do no more than add another disturbing fragment to the already indigestible mix of contradictory forces that rule our lives.

As has been suggested, when reality becomes fragmented, a loss of connectedness, coherence, and interrelatedness of events results. It isn't surprising, therefore, that people in the modern world tend to reverse the logical order of events. We confuse the territorial crowing of the rooster with the laying of the egg, exhibiting accomplishments with achieving them, advertising with manufacturing, and showing with creating. We want to hear the applause before the performance, enjoy the benefits of work before we complete it, spend our money before we earn it.

We can count on this: media personalities will be out in full force in the Age of the Genome to convince their audiences how gene technology should be applied, on terms they will suggest. The mind control of feudalism will return in changed guise in the Age of the Genome, disseminated by means of telecommunication. By sight and sound, media personalities will try to bend the minds of their viewers to conform to their wishes. Many viewers will become unwitting mental captives, television serfs, a "herd of sheep" as Nietzsche would have called them, in spite of the self-determination which the Age of Reason fought hard to win for us. The human genome may assume the mystical aura of the Holy Grail, which in the Middle Ages inspired Christian zeal, King Arthur legends, Celtic myths, and played a role in fertility cults. When people are thus mystified, scientists conducting genetic engineering would be able to discourage legitimate criticism of their goals.

Don Quixote, hero of the book by the Spanish Renaissance novelist Cervantes, drove his spear into windmills in his effort to battle the wrongs of the world. The book highlights the tragedy of idealism throughout human history. Since Cervantes wrote it in 1605, isn't it sensible to ask ourselves: "In spite of exposure to existentialism and to the additional power we have gained from our modern technological breakthroughs, are we coping with an unchanging and unchangeable human nature?" If the answer is yes, many will claim that we can create a better world for humankind only by redesigning our human genome.

The temptation to redesign ourselves genetically may not be easy to resist because the Age of Reason failed to live up to expectations. It is time that we look at the fundamental unit of philosophy to ascertain where we were caught in a trap of our own making. It started when philosophers attempted to define reality.

CHAPTER FIVE

THE DISCOVERY OF REALITY

To enable us to see the future consequences of genetic engineering within a larger framework we now stop to examine some philosophic fundamentals. One of these is basic. It consists of the nature of reality which we shall survey in this chapter.

It is within our perception of reality that genetic engineering will play out its role within the Big Picture view of human life which this book attempts to present. Since an artifically created reality, known as "virtual reality," is increasingly used in experiments involving human beings, an exploration of how concepts of reality affect our thinking is appropriate at this time. We shall discuss virtual reality later in the chapter.

Even before the dawn of civilization, humans probably wondered about a spirit world that seemed neither real nor unreal. Reality might have been questioned long ago when humans scanned the horizon and saw a distant lake that upon investigation turned out to be only a mirage. Perhaps reality perplexed the Neolithic man who at night, after a hunt, dreamed vividly that he had returned to his cave with bison meat—but upon waking the next morning, could not find it.

Much later, when humans were sedentary and could afford to support full-time specialists such as philosophers, they became aware that the concept of reality was elusive and must have argued about what reality really meant. Throughout the ages, debates continued on what reality actually consists of. Contrary to what philosophers would have expected, particle physicists, in the early part of this century, found it necessary to turn to philosophy to account for the unexpected behavior of subatomic particles. Physicists noticed that the tiniest bits of matter were always observed spinning into the direction opposite to the angle from which they were seen, regardless of the position taken by the observer. It didn't make good scientific sense for particles to respond to the way human beings looked at them. Particle physicists concluded that something in the human mind had to be involved in the scenario.

In 1927, the German physicist Werner Heisenberg proposed his astonishing Uncertainty Principle. It entailed the measurement of a quantum system, such as an electron or an atom. It held that any measurement of such a system would disturb it, causing it to become unpredictable. The amount of disturbance was found to be proportional to the precision of the measurement, so that the more precise the measurement the greater the disturbance. The implication was that the laws of atomic physics formulated probabilities instead of certainties, as had previously been assumed. After that, physicists had to conclude that reality was, in part, created subjectively. That notion, of course, put them knee-deep into philosophy. However, I must mention that Gary Taubes entitled a two-page article in *Science* (1994) *Heisenberg's Heirs Exploit Loopholes in His Law*. Probing for reality, whether in the sciences or in philosophy, is a never-ending job. Few concepts have been given so much attention or searched so systematically for hidden meaning by philosophers as the nature of reality.

On the other hand, many down-to-earth people prefer not to think about reality at all. They believe they can live very pleasant lives without worrying about it and they can easily prove they are right. Yet reality does matter in a way not immediately apparent. In fact, reality influences everything people do and that includes even those who are too busy to be concerned about such abstractions as reality. We avoid thinking or doing anything about things we do not accept as real. Nevertheless, unknowingly we all live our lives according to our perceptions of reality. We do so even as we show our children how to kick a football or kiss our mate goodbye when leaving for work.

There can be no sensible discussion of meaning among any group of people unless it is accompanied by an assumed consensus on what constitutes reality. Nevertheless, some philosophers skim over the definition of reality as they expound on less recondite subjects. This causes dissention among them that may have been avoided had they first reached an agreement on the nature of reality. At times, philosophers simply hang the label reality on whatever happens to be in their line of observation.

Since ancient times, philosophers, or lovers of wisdom, have faced different directions and this has left room for arguments. In the 5th century B.C., Plato, one of the best-known philosophers of all times, may have startled his students by claiming that genuine reality couldn't be found in anything that *looked* real. Instead, reality had its own independent existence only in forms or ideas that he called "archetypes of visible things." Things that could be observed by people were merely "shadows" of abstract essences. It wasn't long before one of his famous students, Aristotle, disputed this, asserting that all things people observed were real. He observed that if you stepped on a cat it didn't react at all like a shadow.

The wise men of the East view such musings as delusions. The Hindu spiritual leaders teach their followers that Ultimate Reality is beyond human understanding. The Hindus consider Ultimate Reality, called Brahman, vast and remote. It defies description and can't be seen in the everyday world because it isn't accessible to human senses. One reaches it only through liberation from the cycle of reincarnations by living an ascending order of increasingly blameless lives.

Western scholars have championed their own fragments of reality throughout the ages and given them a variety of names. Depending on their particular focus, they prefix reality with descriptive adjectives such as objective, subjective, empirical, abstract, concrete, physical, sensory, relative, quantum, virtual, primordial, religious, existential, and so on. Although these scholars refer to their fragments as reality, they represent only an assortment of categories and not reality in its larger sense. If a person has one foot planted on a fragment of reality while his other foot rests on a different fragment, he will lose his balance when these fragments drift apart, as they usually do in time. Should he look heavenward instead of down at his feet, he will he forced to invent an innovative philosophy to account for his discomfort.

We should include this possible outcome of fragmentation in textbooks as a warning to future students of human genetic engineering bent on improving human nature. Albert Einstein once claimed even scholars of audacious spirit and fine instinct can be obstructed in the interpretation of facts by philosophical prejudices, stemming, I will add, from the fragmentation of reality. Some people in science and outside of it see only a fragment of a larger whole even though one usually cannot obtain an accurate global impression from a fragment. This is well expressed by the folk saying: "One swallow does not a summer make."

How can we come to grips with reality in everyday life experience is a fair question. Those who search for reality in exotic places resemble people who look for their misplaced glasses while wearing them on their noses. We may fail to recognize reality because it is all around us. It makes some people uncomfortable to think too much about the nature of reality just as many of us get tired of too much discussion on the nature of truth. However, when thinking about the consequences of impending genetic engineering, we must be willing to view reality in its largest context. This means a Big Picture view that includes all the brush marks composing it.

For example, confining our definition of air to that air which escapes from a punctured tire ignores the fact that only a small amount of all the air in the world was pumped into the tire. To be accurate, we must refer not only to the air in the tires on our own car, but also all the air in all the tires in the whole world, as well as to the air not in the tires. This applies to all partial views of reality.

Reality includes everything that exists or has existed. It consists of Plato's archetypes together with all of their "unreal" shadows. It includes every act, thought, dream, scrap of imagination, daydream, hallucination, stick, stone, worm, rat, cat and human being ever born. All things that happen and have happened—all happenings, as I like to call them—are building blocks of which the universe is constructed and constantly reconstructed. In short, everything that happens, seen or unseen, known or not known, creates reality.

Happenings make the universe what it was, is, and will be. Every event, be it a leaf dropping or a star collapsing, share a common heritage with all other things that have occurred. Everything that came into being emerged from the same vast, immeasurable pool of unborn potentialities. In becoming real, they are elite since there is, also, nonpotentiality which can never give birth to reality. The laws of nature permit only some kinds of things to happen in the universe. This gives all things that exist a commonality and even, if one wishes to see it that way, a camaraderie. The potentiality of coming into being relates the past to the present, and the present to the future.

Most of us have the false mental picture of the past as disappeared and gone. The English poet, John Keats sensed this when he wrote: "A thing of beauty is a joy forever . . . it will never pass into nothingness." Continuing in a poetic vein, I agree that all happenings, beautiful or not, leave footprints in the sands of time that cannot be erased. Each unborn potentiality of the universe that is actualized into reality contributes to the ever expanding real content of the universe. The universe consists of an immense parade of events that represent actualized potentialities. Reality incorporates its past within itself and retains it. In preparing the mind for the new age of the human genome, we must become receptive to the idea that heritage is not confined to the history of our family or our nationality or our socio-economic status. Our defective genes, even if replaced, will remain as things in themselves and wonders of nature whether they were good or bad for us. Reality cannot be extinguished, even if billions of years from today the universe collapses into itself and shrivels into a black hole consisting of a featureless radiation that wipes out all past information. On the other hand, if the universe maintains itself in steady state, as some cosmologists believe, past events may become irretrievable. Nevertheless, the reality of past events remains untouched.

All things change, build up and wear down, are transformed, transcended, or metamorphosed. With use, irreducible energy yields to entropy, where it becomes useless for further use. The present yields to the future and becomes the past. Yet neither energy nor reality can be destroyed. One could put this more simply—what is, is; and what happened, both happened and remains. The present and past together make the universe what it will always be. Awareness of this represents a piece of the jigsaw puzzle that must be put in place to construct a new mind-set.

This way of thinking about reality could provide us with an insightful outlook. Let us imagine that a person writes a poem and intends to enter it into a poetry writing contest. Let us assume that the writer somehow lost all copies of her poem. The distraught poet may think that she accomplished nothing by her effort. However, her hope of winning a prize, the challenge, the ideas and feelings expressed in her poem, even the expected admiration of her friends, remain within the reality of her anticipation. An idea seen or unseen by others changes the world simply by having occurred.

Living in our Western culture conditions us to view winning as one side of the coin of achievement, and worthlessness as the other. Both contribute by bringing into reality the potentialities of the universe, that is, making things happen. If one could accept it, this perception could become important to a person, not as denial or as a rationalization, but in a positive way as an alternate view of failure. Gray was mistaken in his famous *Elegy Written in a Country Churchyard*, when he wrote: "the paths of glory lead but to the grave." Instead, the "paths of glory" remain for all time paths along which the universe travels in becoming what it will consist of eternally. Without such a perception of events, the application of genetic engineering may become lopsided and irrelevant.

When it becomes possible to prolong life by means of gene manipulation, people who fear death may form long lines at clinics to have their cell clock-work genetically rearranged to extend their life span. The fear of death has often been ascribed to a fear of the unknown, when actually it represents an anxiety stemming from the seeming end of one's loss of opportunity to continue to actualize the potentialities of the universe. That is why people who feel they have fully used their opportunities to actualize potentialities of the universe believe, as they might say, they "made a difference" in their lives. They are able to accept death more readily than others. Lao-tsu wrote in his book *Tao Te Ching* in 640 B.C., "A man is ready for sleep after a good day's work." This view permits us to feel that we can gain satisfactions without fame or recognition, merely because we increased the world's reality by what we did. Every act is a creation. That is why the death of a young person who did not yet have much opportunity for self-actualization strikes us as a greater tragedy than the death of an older person, who had greater opportunity to actualize some of the potentialities of the universe.

The influential 16th century French philosopher Descartes may not have agreed with me. He did not accept any happenings as reality without proof and even questioned the existence of any reality at all. Like many others, he was convinced that events had to be demonstrated tangibly to become real. This view of reality is the extreme opposite of the one I have described. Finally, Descartes felt compelled to search for evidence of his own reality. Reluctantly, it seemed, he had to conclude that his ability to think proved that he existed.

Some present-day philosophers believe that by turning to the neuroanatomy of the brain they can locate the origins of philosophical truths. They try to find the sources of human consciousness within neurons, synapses, and neural networks. But they will not find consciousness locked away within anatomy. Physiology relates to consciousness in the same way that existing does to thinking. Descartes would have said, "I am conscious; therefore, I am alive," when he should have said,

"I am conscious and I am also alive." If everything that exists represents reality, as I have maintained, then a lack of consciousness would also have some kind of an existence.

Reality is sometimes confused with truth. However, both truths and untruths create reality. A statement in a book by two psychologists, Grant and Evens, reflects a common error. The authors write, "We are what we do . . . it is our behavior that defines what kind of people we are." What they fail to say is that reality is not created by action alone. Inaction creates reality as well. If a man digs a hole using a shovel, he creates reality by what he does. If another man with shovel in hand simply stares at the ground, he creates reality by what he doesn't do. In each case something that could occur did occur, and occurrences make reality. This holds also for merely finding joy in living and appreciating life.

An unusual creature called a Tardigrade can exist for more than a hundred years without water, oxygen or heat in a state which by most definitions would be called death. Yet when moistened it immediately springs back to life. Some 1 billion-year-old fossils of blue-green bacteria and their modern counterparts which form a living film on stagnant water look exactly alike. The survival of this species without apparent changes over such a length of time actualizes a potentiality of the universe, as do new species that undergo rapid changes.

In psychiatry, schizophrenic patients are described as having lost contact with reality. The description serves to communicate the state of the patient's mind but taken literally reflects the prevailing misconception of reality. One cannot lose contact with reality. Instead, the patients' hallucinations and delusions represent realities to which their doctors do not subscribe.

To illustrate this, I offer an analogy that certainly seems ridiculous. In a hypothetical scenario, one can imagine one is eating a meal and consider it as real as actually putting food into one's mouth and swallowing it. Certainly, merely imagining that one is eating a meal would be of no satisfaction to a hungry person. Regretfully, in this world, satisfaction and reality often remain independent of each other. No one would feel less hungry if told that fantasizing eating a meal activates more cortical neurons in higher brain centers than actual eating does. Nevertheless, both of these alternatives create reality even if one of them would eventually lead to starvation.

When we hear something described as "not real" we should be aware that this is impossible, since everything that exists represents reality. This perception of reality is highlighted by the discovery that has been called virtual reality. The American Psychological Association monthly publication *APA Monitor* (Vol 27, Number 3, March 1996) headlined its front page with, "Psychologists Dive Into Virtual Reality." The report states: "Instead of being bystanders, we'll be active participants in worlds that we can control and interact with from the comfort of our recliners." It continues: "Virtual reality puts people into computer-generated worlds that can make them feel real enough to cause them to sweat, shake, and sway." Some computerized systems project images of a virtual reality world onto a video-screen mounted inside a helmet that a viewer wears. The *Monitor* explains that under these conditions viewers then see images from the same perspective as they see the world itself. This allows them to measure and respond to items in the virtual world as they do in the real world.

As in genetic engineering, hazards exist in manipulating human beings by means of virtual reality. Another article in the same issue of the *Monitor* warns that virtual reality may result in motion sickness and, beyond that, have a "lasting affect on a viewers perception." This would create a new human malady called "cybersickness." According to the *Monitor,* one conclusion is that "the penalty for fooling mother nature can be a set of symptoms that are on one hand unpleasant and on the other hand insidious." Some researchers speculate that a person could cause a traffic accident while adjusting spatially after spending time in an virtual environment. Another example of the reality of "non-real" things can be taken from medical research. The sugar-filled pills we call placebos have worked remarkably well in reducing many symptoms of diseases and have also created side effects that only genuine medications are supposed to cause. Although a placebo and a proven medication are very different in content, our emotions and physiology may respond to both as real.

Let us take this idea just a bit further. Over thirty years ago, a popular writer on human nature, J. H. van den Berg, noted in *The Changing Nature Of Man*: "It is quiet clear that geography does not play a role in the world of a person who says 'I shall be with you tomorrow evening' and the person to whom he says this happens to be on a train that departs for Switzerland in a few minutes. And yet he is not lying." Van den Berg points out that she (the person to whom the above was said) knows that the man who said it is thinking of her. The author describes her as becoming quiet as she feels his imaginary presence. The author continues: "Everyone has experienced that a loved one is with one in spirit . . . This is the secret of all thinking, and of all longing."

Let us now contrast van den Berg's view of reality with that of T. S. Eliot's in *Murder in the Cathedral* (1935). He describes four "tempters" who lament: "Man's life is a cheat and a disappointment. All things are unreal. All things become less real; man passes from unreality to unreality." The tempters suffered from a common misconception. Their "unreality" is one of the most poignant aspect of reality. Hopefully, future neurogeneticists will accept the idea that our most human qualities come from that element of the unreal that represents the greater-than-the-sum of our anatomy. Genetic engineering has been likened to a "magic bullet," but human behavior more closely resembles the movements of a ping-pong ball batted back and forth in a game played by amateurs than to the trajectory of a bullet fired at a target by an expert marksman.

In making momentous decisions, considerations must be given to the reality of things as they are not, as well as to things as they are. Human behavior is rooted in imagination, daydreams, and fantasies, as well as in events that are concrete and objective. The concrete thinking, here-and-now mind alone will never grasp the significance of the important abstractions. Our own modern culture undervalues daydreams. A society that values deeds too highly and thoughts too little produces people with superficial satisfactions and underlying discontentments. Could this explain Thoreau's often quoted observation: "The mass of men lead lives of quiet desperation." Now is the time to ask, will someone in the future try to alter with gene replacement the desperation that our culture—not our genes—has given us?

We cannot know the universe in its totality, because it represents an abstraction that our cognitive capacity cannot grasp. If we reduce its essence to the level of our subjective perception, it allows us only partial views. However, there is something akin to the "beyondism" of Ultimate Reality within us. It is precious because it enables us to identify with the universe—experience it—in spite of our limitations.

Genetic engineering represents a biological technique and, in that respect, it is a one-dimensional approach to the multi-dimensional human beings it intends to serve. The danger exists that in planning to modify the human genome, fragments of reality may masquerade as the whole of reality. Should the pretensions of any of these fragments beguile those empowered to apply recombinant gene technology to humans, they will fail to see that reality extends beyond the coded messages on our DNA.

Scientists must often concentrate on a fragment of reality in order to discover revealing aspects of the whole picture. They are, however, aware that the fragment is not the whole. Under some circumstances there are justifications in ignoring the Big Picture. For the sake of their mental health, some people may deliberately narrow their notion of reality. It could be therapeutic for them to exclude the turmoil of the world and focus exclusively on a tiny, comforting and familiar fragment of reality. A single loved one, a mate, child, pet, or even a material object of sentimental value would then become their exclusive focus. There are times in life when a turtle must pull its head in under its shell. Such situations represent a withdrawal from reality rather than a fragmentation of it.

If we develop a mind-set in which we take responsibility for shaping the universe, our individual lives will take on a new dimension. I have put this thought into a verse in my book *Poems For Living* (1995):

> All things are eternal.
> Nothing disappears,
> Your hopes, your deeds, your daydreams,
> Your laughter and your tears.
>
> Your deepest thoughts and secrets
> Through light-years will not fade
> Because they became building blocks
> Of which the world is made.

In the next chapter, we shall examine ideas that deal with morality. Being moral cannot be viewed as floating rudderless through life. It must be seen as anchored into the reality on which its existence depends. If we are aware of the fact that reality cannot be fragmented we shall realize that morality, also, cannot be fragmented without loss of its meaning. Morality plays a leading role at a time when the ethical applications of gene technology and of virtual reality offer us the greatest challenge we have confronted since we evolved as a species.

CHAPTER SIX

A MORALITY TO FIT THE AGE OF GENETIC ENGINEERING

Our mind-set shapes our opinions about events that occur in the world. Our opinions, in turn, lead us to the kind of morality we adopt. If we apply genetic engineering before we commit ourselves to a human-centered morality, gene technology will shape our ethics to its own ends. Therefore, a commitment to a morality appropriate to gene technology must precede mass application of genetic engineering.

Today, morality has the characteristics of a conundrum. What are the qualities of morality that make it difficult to reach consensus on its meaning? To consider this question, we must realize that from the beginning of recorded history people have lived in separate groups, families, clans, and tribes. Later, different religions and nationalities evolved. These caused the creation of dissimilar cultures, some with contradictory world views. From these, morality emerged.

We are able to perceive abstractions such as virtue, goodness, and fairness. Had these perceptions been strongly reinforced by the world's rulers and political leaders, they may have brought about an inclination for all peoples to live together peacefully and amicably. Some sociobiologists postulate that morality is derived from an instinct to preserve our species. In the next chapter, I shall explain why I believe this view represents an oversimplification.

Our longing for a brotherhood and sisterhood of the world's peoples could contribute to the survival of our species. These longings go beyond a struggle for species survival. Granted that the instinct of self-preservation, for the individual as well as for the species, is a powerful one. However, cognition, idealism, and even spirituality are also involved in the longing for peace on earth. We must try to make sense out of the perplexing observation that the human dream for peace on earth has never been realized for any appreciable length of time in recorded history. Almost everyone has ready answers to explains this failure. Regardless of religious, psychological and sociological explanations, the fact remains that we have experienced on-going mutual enmity and even the denial of another group's humanness. Our feelings of affiliation and acceptance of each other as human beings have been limited and transitory. Even the ancient Greeks, who developed one of the most enlightened civilizations, referred to those who were not Greek as barbarians. Strife, rather than harmony, has ruled much of human life.

We cannot put this unhappy situation at the door of civilization. Before there was what we call civilization, men competed for hunting grounds and sponsored raiding parties to wrest women from neighboring tribes as well as to relieve boredom. We have heard fatalistic explanations for our inability to get along— "human nature is like that." If so, we have no business meddling with the basic building blocks of life. If wars are inevitable, then there is no doubt that some time in the future genetic engineering will become a tool of war, just as almost everything else suitable for this purpose has.

The world's major religions acknowledge our oneness as a human family. Isaiah, a prophet of the Old Testament, proclaimed in the 6th century B.C. that the God of Israel was also the God of all people. The Judeo-Christian tradition recognized the unity of all peoples through their common creator. Christians believe that Jesus died for all of humankind. According to Islam, God created the universe out of mercy for all people. The Koran declared that making the earth livable for all the world's people is an ideal endeavor. Buddha attacked one of the great barriers to human unity—the idea of inherited elitism. Since class differences separate people, Buddha taught that a person's birth into a family of high rank does not increase the person's spiritual worth.

In contrast, some biologists view intraspecies competition as an indisputable fact for all life, including our own human life. Animals have intragroup struggles for dominance, but they do not have a conception of a higher supernatural power. Whether or not we believe in the supernatural is not the point. We are capable of conceiving of something supernatural, and this suggests that we can evolve a morality that lies beyond the natural limits of other species. An awareness of our common genetic heritage provides biological support for the concept of the oneness of humankind. This leads to the conclusion that all members of our species have a joint ownership of our human family property—our genome. In order to handle the responsibility of redesigning life, we must find a way to minimize hostilities and jealousies that from time to time have split our species into fragments. If we hope to avoid using gene technology against each other, we must rely on the kind of morality that will bring the fragments together. This can be brought about more easily if we do not fragment our concept of reality as suggested in the previous chapter.

In the past, only military conquest could hold large population groups together as Rome did in Europe and the Incas did in South America. Military conquest as well as religious missionaries often spread their unrelated cultures and religions to distant areas. A morality suitable for the age of genetic engineering must replace our vague awareness of our human relatedness with a strong focus on our oneness. Only then shall we be able to reach a level of consensus upon which we must rely upon as our guide in applying molecular biology. Before the human genome project is completed, a new sense of human unity must be discovered and put into practice.

I have said earlier that before the advent of civilization people were not very different than they are today. The environmental situation of hunting and foraging societies did not demand that they establish territories nor compete for dominance. It was the circumstances, not the people, that differed. The 18th century philosopher Jean Jacques Rousseau claimed that indigenous peoples with simple life-styles were moral in their natural state. He was mistaken. He failed to contrast their non-competitive life styles and mutual dependency with our later hierarchical and competitive societies. Without environmental pressure for doing so, they had little reason to exploit each other.

After the founding of the city-states, people began to discuss right and wrong behavior and to argue about how to define it. Susan Kent, Professor of Anthropology at Old Dominion University, did extensive studies with the indigenous people in Botswana, Africa. She wrote some of her observations in *Farmers as Hunters—The Implication of Sedentism* (1989). Kent recognized that sedentism (residing in one location) created a new potential for interpersonal conflict that cannot be resolved as readily as when groups are mobile. She explained that after a population becomes sedentary "fissioning and moving to different areas [to escape conflicts] are no longer viable options." Thereafter, "people with restricted mobility must address intragroup conflicts in other ways than through movement." Conquests and wars have become popular alternatives.

With our hugely expanded population, how can we regain the communal sense of belonging to each other that we have lost through sedentism? It will be difficult, since hierarchies and other divisions of people into classes, tribes, and nations serve to stabilize societies internally. Victories and defeats among people became one of the focal points around which a group's traditions are formed. Traditions provide people with a strong wish to maintain them. It can be put into simple language: We are in a rut. Therefore, creating a new mind-set has to overcome formidable obstacles.

Before the world's major religions evolved, traditions were passed on from generation to generation by means of oral legends. These didn't try to bring about world-wide consensus, because they were designed to explain the unknown, to teach and to establish traditions applicable to the culture of a specific group of people. This limited role made a concept of a universal morality difficult to grasp. Therefore, many people willingly accepted the definitions of morality provided by their own religious leaders and philosophers. However, Plato, in *The Republic, Book II*, maintained that "unless a person, himself, is able to abstract and define rationally the idea of morality . . . he apprehends only a shadow if anything at all." Plato thought that philosophers were best equipped to see beyond the shadows but that individuals must also work out the nature of morality themselves.

Modern religions differ in some respects in regard to what should be considered right or wrong. Still, there is a common thread that runs through the definitions of desirable behavior such as the Golden Rule. Christianity, Brahmanism, Buddhism, Judaism, Confucianism, Taoism, Zoroastrianism, and Islam agree using different words: "Whatsoever you would want that people do for you (that is helpful and good) do this even for others." This failed to be universally accepted as a guide to action because people differ among themselves about the specific applications of the Golden Rule. It also failed because people felt no need to adhere to it.

Throughout the years, people continue to disagree about what morality consists of. Some have accepted hedonism, others adopted situational ethics or pragmatism as valid bases of morality. Sociologists and some anthropologists continue to view morality as a product of a people's culture. Ethnological studies reveal that perceptions of right and wrong in different societies frequently are not in agreement. This convinces some people that ethics and morality can only have situational applications. This view makes morality dependent on a situation, instead of the situation dependent on a morality. Thinking this way could lead us into deep trouble should we apply it to genetic engineering.

I have already mentioned another erroneous interpretation of morality—adaptive cultural evolution. This view represents no more than the nurture side of the nature/nurture controversy. The idea that human morality resides in the structure and chemistry of the brain is a fallacy. It assigns the locus of human endowment to what the head contains. This view shares a basic misunderstanding of human attributes with phrenology. The discredited theory of phrenology was proposed by Lombroso, the 19th century Italian penologist, and identified human traits by interpreting bumps and bulges on the skull. Lombroso's false assumptions do not differ in principle from the claims of those neuroscientists who view the origins of behavioral traits as residing in the brain.

It is true that brain lesions, toxins, and electro-chemical stimulation of certain parts of the brain can produce fears, anxieties, and hallucinations. Nor is there doubt that chemical changes, structural differences, amounts of serotonin, dopamine, and other enzymes in the brain significantly affect human behavior. We know that abnormal brain conditions can turn good-natured persons into violent ones who lose their judgment and their capacity to distinguish right from wrong. However, in no way does this prove that morality originates in the brain. A radio whose quality of tone is distorted by defective internal components does not prove thereby that it creates the sound it emits within itself. Instead, brain defects and defective radio parts share an inability to respond to external transmitting stations. We must understand that morality is first of all a potentiality of the universe that was brought into reality by human choice. Morality viewed as originating and residing within human physiology represents a fragmented view of reality. This does not lessen our admiration for the human brain which has correctly been called one of the most remarkable structures in the known universe. Pre-existing potentiality is what we, who are scientists, are apt to ignore to our own disadvantage. Doing so distorts the Big Picture view and leads to the small picture view of reductionism.

Long ago, Newton's reductionism, the notion that cause and effect explains all of the universe's phenomena, gave way to the new physics of quantum mechanics. It became further out-of-date with Heisenberg's principle of indeterminacy and the emergence of a science of chaos, mentioned earlier in this book. However, in the behavioral and social sciences reductionism still plays a role in the attempt to account for abstractions like morality. Scientists are uncomfortable with the idea that genuine human morality is not in accord with how we previously viewed nature. Morality represents a human actualization of a potentiality that, because it lacks precedence, makes defining it difficult and open to conflicting interpretations.

Two centuries ago, the German philosopher Immanuel Kant defined morality as "conduct that requires that nothing be gained by it." As the Greek philosopher Diogenes showed in the 5th century B.C., most people do not act to gain nothing; they have ulterior motives. Long before him, the Greek philosopher Diogenes, according to legends, walked through the streets of Athens with a lighted lamp, looking for an honest man. He could not find one. It seems that his widely publicized search was conducted to demonstrate that those who professed to be honest had ulterior motives. Their "honesty" was insincere and did not fit his definition of honesty. We must admit that if Diogenes did his research today the results would be similar.

A commitment to a morality that, in Kant's view, doesn't offer rewards seems almost as amazing as the discovery of genetic engineering itself. Kant's type of morality cannot serve adaptation if it consists of behavior from which nothing is gained. It seems incredible that the idea of morality retains meaning only when it provides no advantage to those who commit themselves to it. Yet its inability to serve as a means to an end remains an inflexible criterion of its genuineness. People cannot convert morality into something other than that which it stands for, or reshape it to make it acceptable to themselves. Morality does not bend in the wind. It can only break into meaningless pieces. This leaves us no choice but to accept the proverbial saying found in folk wisdom: "Good deeds are their own reward." This is precisely the kind of morality we require to make genetic engineering safe for humankind.

The human penchant to anthropomorphize inclines us to give humanlike qualities to our external world. A century ago the noted German philosopher Hegel made an ultimate projection when he declared, "Man is the universe in miniature." Against all of this projection, morality, as Kant described it, stands out as a towering exception. Nothing of ourselves can ever rub off onto this kind of a morality. Some of the content of such a morality can only rub off on us. There is no reciprocity. Because it does not grow out of our current mind-set, people who reflect on this aspect of morality may find their confusion about its unusual characteristics increasing. It may make it easier for us who are now living in a reward/punishment world to conceive of "reward" as bringing into existence something merely for it to exist. We see traces of this among some people who wish to keep an endangered species alive just so that it continues to remain part of the world, without necessarily helping us to survive.

In the middle of this century, J. Bronowski, a distinguished scientist and philosopher formerly at the Salk Institute for Biological Studies, in *Science And Human Values* (1965), attempted to clarify the paradoxical nature of morality. He pointed out that morality does two different things at one time. It brings people together into societies and also requires that each person within a society remain an independent individual. Bronowski came to the conclusion that any ethic that fails to accomplish both of these requirements could neither evolve workable and desirable values or even permit them to exist. Without them, ethical convictions are likely to lead to prejudices and doing what is right would deteriorate into self-righteousness.

Those who control experimentation with genetic engineering must take into account that morality which demands freedom also limits it. Moral persons are bound by their own freely chosen rules of ethical behavior. Once they make a commitment to a set of values, they must adhere to them unless they change their perception of morality. But this must also be a moral decision and not one chosen for gain. In the context of genetic engineering, we should take into account that morality which demands freedom for the individual also limits it.

In the next century, gene exchanges between various forms of life will be created more easily than at present. Then it becomes critical that we observe the limitations of behavior that Kant's kind of morality imposes on us. More than ever, in the approaching Age of the Genome, we shall need a morality to guide us that meets Kant's and Bronowski's criteria.

Some prominent philosophers ignored the Kantian view. Both William James, a pragmatist living in the 19th century and John Dewey, the American educator who lived until the middle of the 20th century, promoted the idea of morality as a means to an end for achieving beneficial social consequences. We have already noted that it makes sense to view morality as something that exists for the benefit of society. However, this use leaves the door to morality open equally to humanitarians and bigots. Each of these could use "morality" differently to support their causes.

Those who demystify morality and reduce it to the level of a tool fail. They fail to realize that in doing so they deprive it of its unique asset—independence from serving as a means to an end. The writers of the Judeo-Christian Bible confronted that problem in the Old Testament Book of Job. The illness and grief that befell Job taught him the lesson that good fortune isn't necessarily a reward for ethical behavior. Job did nothing to deserve the suffering he was forced to endure. The purpose of his lesson was to demonstrate that virtue must be pursued for its own sake. When the lesson was learned, the Bible tell us that the Lord blessed Job with 14,000 sheep, 6,000 camels, 7 sons and 3 beautiful daughters; and Job enjoyed these blessings until his death at the age of 140.

The story has a happy ending. The writers of the Book of Job shrewdly judged human nature. People at the time it was written, as today, could not believe that good behavior would not be rewarded. After Job accepted morality for its own sake, his fortunes had to improve or few would wish to learn this lesson. Nevertheless, the point made is that Job did not anticipate his reward and it was not his goal.

The story of Job illustrates the genuine meaning of morality as a voluntary commitment, adopted for its own sake without expectations of rewards. Only human beings are capable of responding to behavior that seems to disregard the laws of the pain/pleasure principle, which is that all actions are elicited solely by the pleasure they create and the pain they avoid. The capability to override these laws enables us to experience love, loyalty, and respect for law without having ulterior motives as incentives.

The trial of Socrates is a famous example of a moral commitment to the law. At his trial in Athens in 399 B.C., Socrates was accused of introducing new divinities and teaching rebellious ideas to the Athenian youth. He was found guilty and condemned to death. His friends offered to help him escape from prison, but he refused their aid. He accepted the court's sentence and drank the hemlock, which at that time was used for capital punishment. His commitment to morality required him to obey the existing law because of principle, regardless of the consequences to him personally. His was a cognitive decision that overrode his instincts for self-preservation. Where else but in a non-utilitarian human world could anyone say sincerely, "it is better to give than to receive?"

In the Age of the Genome, more than at any other time in human history, people must give up the freedom of indiscriminate action and assume the responsibility required to make this a moral world. Herein lies the key that could protect the human genome from exploitation. As stated previously, because morality restricts freedom and cannot serve as a means to an end, it does not allow individuals to become victims of unfair group-centered social designs. If those in power in an age when gene technology can be applied to humans do not adopt Kant's definition of morality, they might mandate the genetic engineering of others solely to gain control of them. Those controlled would lose their free will. Obviously, individual freedom cannot exist among people genetically designed to have no alternatives.

Today, many intelligent people believe that moral behavior as a thing-in-itself without advantages belongs to a kind of moral science fiction. They feel that moral behavior offers advantages that idealists obscure because they are unaware of them or do not wish to acknowledge them. Perhaps that is why the Greek hedonist Aristippus, in the 4th or 5th century B.C., proposed a concept of morality more credible to some people. He proclaimed enjoyment and physical pleasure as the highest of virtues. Throughout the millennia, Aristippus has had no lack of followers. There will be plenty of people in the Age of the Genome who will heartily agree with Aristippus. Genetic engineering may admirably suit their purposes. Today, hedonism is wearing new clothes and called utilitarianism or modern Darwinism.

I have previously suggested that in the future some people may look forward to the day when we shall identify the combination of genes that activate the neurotransmitters, serotonin, dopamine, and other euphoria-creating enzymes. Thereafter, hedonists might gain lasting genetic "happiness" with less effort than it takes to obtain the common street drugs in use today. Nothing that previously existed could produce permanent enjoyment as efficiently as genetic engineering, once the human genome's pleasure-producing characteristics are mapped and molecular biologist learn how to activate them.

The French philosopher, Henri Bergson (1911), a Nobel Prize laureate, coined the term "creative evolution" to describe developments that occur independently in the human mind. Bergson's "creative evolution" allows for a touch of craziness within us that those who expect humans always to act rationally fail to take into account. Perhaps a recent view of our touch of craziness puts it in more acceptable terms. Researchers discovered that chaos exists normally in the human brain. Because it offers an alternative to "every man has his price," our touch of craziness leads us to our greatest innovation—behavior valued as a thing-in-itself. It enables us to adopt morality simply because we want to be moral persons. This isn't natural, logical, or rational, and that is why morality still remains a subject of controversy and confusion. However, we should expect that in the history of the universe unprecedented developments can transcend the past and lead life into new directions, both in nature and in human nature. This gives us hope that we can develop a new mind-set.

Life on earth evolved about three billion years ago and was an innovation, in that life was able to replicate itself and to struggle to survive. Our human ability to conceive of abstractions such as a non-opportunistic morality represents an innovation of similar magnitude. Because it seems to override the opportunistic laws of biological nature, many people feel that we shall be unable to achieve such a seeming miracle as morality valued only for itself. Therefore we expect supernatural powers to come to our rescue, forgetting the folk wisdom that reminds us that "God helps those that help themselves." We can help ourselves by furthering creative evolution through environmental and cultural changes that would lead to the adoption of genuine moral values and new mind-sets. This would occur in a society that exposes its children to inspirational materials involving morality and ethics.

Our dilemma, in part, may be traced to our shift in self-evaluation. In the Judeo-Christian tradition, we were told that we were "only a little lower than the angels." Our contemporary re-evaluation places humans as, "only a little higher than the animals." From "almost angels" to "almost animals" is a long downward slide. It represents a change of view that is bound to create the equivalent of a bad culture shock. Worse than that, it has become a self-fulfilling prophecy, which has crystallized to become a tradition that we live by.

In the next chapter, we may find that the new mind-set necessary for the age of genetic engineering lies closer to the biblical human-angel designation than the contemporary human-animal one. I say this as a scientist who does not believe in angels and who is very much aware of the intelligence of animals, as well as an admirer of the astounding adaptability of insects.

CHAPTER SEVEN

THE FALLACY OF HUMAN/ANIMAL COMPARISONS

Two court trials separated by more than two thousand years stand out like beacons in the history of Western civilization. One of these is a trial held in Athens, Greece, in the year 399 B.C. and the other took place in A.D. 1925 in Dayton, Tennessee.

The accusations against the defendants in these trials were strikingly similar. Socrates was charged with violating the laws of Athens by teaching the youths of his city religious heresies that offended the gods. John T. Scopes, a high school biology teacher in Dayton, was accused of violating the laws of Tennessee because he taught public school children the theory of evolution, which contradicted the Bible. Among those who objected to the teaching of evolution in 1925 were religious people who believed that a theory that contradicted the Bible was blasphemy. Others were against teaching evolution because they believed that it reduced human worth by viewing them as no more than intelligent animals. Socrates was found guilty and condemned to die. Scopes, defended by Clarence Darrow, one of America's leading criminal lawyers, was also found guilty (the verdict, however, was later reversed on technical grounds). Scopes' trial received world-wide publicity. The press called it the "Monkey Trial" because popular belief at the time was that evolution meant that humans were descended from monkeys. Evolution, however, does not suggest that humans are descended from modern monkeys or apes. The first pre-humans branched off from the primate tree about three million years ago. Monkeys and, later, apes were different and earlier branches of the tree. This chapter attempts to show that viewing humans simply as intelligent animals is flawed. This does not imply that animals are "lower" than humans. We differ from animals in significant ways that are not immediately apparent. This difference is sometimes glossed over by those who think that putting humans on the level of animals is democratic. Many sensitive and insightful people revolt against the idea of human superiority and stress the oneness of life and nature. I agree with them. However, pointing out how humans differ from animals does not imply superiority. There is no need to put either arrogance nor humility into the picture. Our human qualities are unique, but we are not an aberration in nature. Unique systems are common in the animal kingdom. Think of sonar bats, the celestial navigation of migratory birds, or even the unique trunks of elephants. We have the capacity for spoken language, but we can not hoot like chimps nor communicate by waggling like bees do.

Some biologists and zoologists and especially sociobiologists tend to underplay significant distinctions between humans and other animals. The abstractions we are capable of making and concepts like morality make us as different from other animals as animals are from plants. We make tools as some animals do, but only humans use their minds to break up configurations like trees and visualize in them potential bows and arrows, canoes, axe handles, firewood, and lumber to construct houses. Some animals also have the ability to make tools, but only humans use tools to make other tools. As a rule, scientists are not philosophers and they tend to focus on genes, brain structure, blood types, intelligence, and physiology. No one can deny the physiological and psychological similarities between humans and intelligent animals, especially apes and dolphins. Nevertheless, the chasm that separates us from all other forms of life is enormous—for instance, we attend church!

In itself, attending church actually has nothing to do the point I am attempting to make. I refer to the fact that all animals lack the capacity to imagine anything like a house of worship. Only humans can ponder their origins and think religious and philosophical thoughts. We alone ask the non-productive "why" questions of our existence which contrast with the simple utilitarian "how" questions of obtaining food and minimizing pain. The difference between humans' "why" questions and other creature's exclusive "how" questions is unbridgeable. We alone exert ourselves to bring potentialities out of the shadows of the universe merely to create reality and for no other reason. Only humans strive to bring something into being simply because we want it to exist. We strive for survival of something beyond ourselves and our species. This is a breakthrough. Like other forms of life, the pain and pleasure principle rules us to a large extent, but not exclusively. We can go beyond the small picture view of our existence and attempt to grasp the Big Picture view that must include the entire universe. The activities of our astronomers and cosmologists prove the point.

Modern writers continue to call our attention to human/animal similarities. In doing so, they find identical qualities which lead them to believe that both humans and animals are biologically endowed with a capacity for "genuine love." It is not difficult to see the signs of sympathy and caring among animals. Observers of animals in their native habitats, as well as owners of pets (such as my own family; we've had several dogs, cats, a monkey, turtles, tropical fish and a canary), have seen cooperation, sharing, and grieving among animals. Observers have also noticed that chimpanzees and other primates resemble us in many important ways. Nevertheless, false conclusions are often drawn from such observations. They include the idea that human unselfishness is governed by the laws of the survival of the fittest and exists for no other reason than that natural selection favors it.

To illustrate our capacity to consider the Big Picture view, I shall quote the last four lines of a poem written by a friend of ours. He had spent many years at sea as a navigator. In his poem, he identified himself with the sea and asked,

> Am I a strong, uncaring sea which does whatever it wishes,
> Or am I a fickle sea moved only by wind and current?
> Are the moods and tantrums of the sea malicious,
> Or are they the unexplained vagaries of something beyond?

Within these kinds of questions we can discern the roots of a new mindset.

First, however, we must respond to the arguments of those who deny our uniqueness and prefer to believe that our innovative morality is shared by other forms of life. In an age when we possess the power of redesigning life itself, it is important to examine this matter further. It is necessary that we carefully investigate claims that we are merely intelligent animals. Our capacities contradict this view. If we are more than just intelligent animals, many people believe we have responsibilities to the world that we cannot share with non-human forms of life.

There are some sociobiologists who see human altruism as no more than instincts that cause some animals and insects to sacrifice themselves for the survival of their species. One proponent of this idea is the evolutionary biologist David P. Barash, professor of zoology, University of Washington, who wrote with enthusiasm on this subject. In *Sociology and Behavior* (1977), he maintained that "Sociobiology is a whole new way of looking at behavior. It is the application of evolutionary biology to social behavior that may hold the promise for a greater understanding of human behavior." If this were true, it would greatly simplify the concept of morality. Unfortunately, the opposite is the case. Barash's belief in the animal origins of morality muddle the concepts of human behavior even further. One can often find animal acts that have the appearance of morality. Rattlesnakes never bite each other while fighting. Antelopes refrain from using their potentially lethal horns against each other. Deer lock horns in battles to establish dominance but don't use them to seriously injure or kill one another. Some termites commit suicide by exploding their guts over enemies who threaten their nests. Rodents who contract contagious diseases have been known to starve themselves to avoid infecting the others in their burrow.

The temptation to see human morality as derivative of these types of animal behavior is great but, nevertheless, false. Our enormously greater range of behavioral options distinguishes us from other forms of life. Human imagination can override any of our instincts that would prohibit us from killing another human being. We only need to read our history books or the daily newspapers to know that in our imagination we can dehumanize people and put aside any inhibitions we might have against killing members of our own species. Yet our killing each other triggers another human quality that is unique. It is our ability to create the fiction that allows each side of a conflict to justify waging wars in the name of morality.

Some sociobiologists believe that humans can attain peace on earth only after they meld into one super-organism analogous to beehives or ant colonies. Edward O. Wilson wrote in *Sociobiology* (1980): "a single strand does, indeed, run from the conduct of termite colonies and turkey brotherhood to the social behavior of man." Such biological-centered views differ substantially from the scenario prescribed by the world's spiritual leaders, who exhort us to live peacefully with each other in a world community. Their emphasis is not utilitarian, but merely that this would be the right thing to do. The concept of community entails cognition and choice, while the idea of a super-organism implies automatism. Insects are drawn to flowers solely for the ulterior motives of feeding on their nectar, and flowers attract insects simply in order to scatter their seed. Only humans, who believe in beauty for its own sake, are attracted to flowers for their loveliness, rather than their utility.

Yet the argument against human uniqueness continues. Those who disagree with the view that humans are extraordinary in nature can produce impressive anecdotal evidence. To demonstrate that we are merely linearly more advanced than animals, observers of animal behavior point out that animals are capable of solving problems, planning, even deceiving other animals and humans. This ability requires insight into one's own behavior, as well as knowing how it will effect the mind of other creatures. The media have carried stories about chimpanzees who are smart enough to search for rocks sufficiently hard to crack their favorite nuts. They also strip twigs and use them to extract termites from their nests. A chimpanzee and a gorilla living in homes with humans attained fame by media reports of their use of the American Sign Language to communicate with their keepers. Green-backed herons in Japan were observed breaking off bits of wood and then dropping them into bodies of water to catch minnows attracted by the bait. A Capuchin monkey was observed applying compresses of leaves and mud to the wounded head of her baby. Who can deny that these activities resemble human traits and abilities rather than our stereotyped ideas of animal behavior?

Lewis Thomas, (1913–1993), a medical educator and writer, observed that, "Ants are so much like human beings as to be an embarrassment." He mentions that they raise aphids as livestock, farm fungi, use chemical sprays to alarm and confuse enemies, launch armies that wage wars, and capture slaves. Thomas concludes, "They exchange information ceaselessly. They do everything we do but watch television." However, our canary watches television with our family. Thomas failed to mention that he did not see ants attending church nor give any indication of knowing what a house of worship might represent. No need, therefore to feel embarrassed yet. Ants do not resemble us now. But we may have reasons to be embarrassed later. Ants would closely resemble the species that may replace us—*Homo geneticus*—unless we refrain from genetically engineering people until we are morally prepared for it. We can be sure that *Homo geneticus* would find embarrassment no lasting problem. The species could genetically engineer itself to be unable to feel embarrassment.

Dogs were bred from wolves who had instincts to protect their pack. Dogs who risk their own safety to protect their masters from harm act from instinct which can be reinforced by training. Such so-called "moral" acts are different from those correctly called "morality," although they resemble them. Likewise we resemble apes in blood types, intelligence, ability to recognize signs and symbols and even, possibly, a sense of humor. These similarities tend to obscure our vast differences from them. The principles of morality proposed by the Golden Rule involve cognition and spiritual concerns. They differ fundamentally from any behavior of animals, whose actions are triggered by species survival instincts or social conditioning. In spite of their biological, physiological, and to some extent psychological similarity to us, they lack the capability to adopt a freely chosen morality. If we find conclusive evidence that certain animals do voluntarily commit themselves to a morality not prompted by instincts and that offers them no advantages, we are obligated to consider them eligible for public school education, social security, and unemployment benefits.

Finally, I will point out that no species, including our own, is a match for the cleverness of viruses. They are far ahead of us in genetic engineering. They possess only DNA or RNA which serves as a pattern for duplicating DNA. They use other forms of life—humans are popular—to create the DNA they need to reproduce. Viruses enter living cells with the equivalent of a key and, once inside, use the occupied cell's DNA to reproduce their own kind. It will take much more than our present cunning to outwit them. No, we humans have no reason to feel superior. Nor will *Homo geneticus* have a reason to feel superior to us, even if they would exhibit a few remaining humans in zoos.

Evolution spells out success in terms of reproduction of offsprings which, in turn, are capable of producing offsprings. Using these criteria, insects are far more successful than we are in surviving and adapting. Some people suggest—science fiction writers among them—that, given enough time, animals may undergo mutations that will make them into super-animals. With more complex brains than ours, they might capture us and keep us in their homes as pets, like we do cats and dogs. Even this would not matter in terms of what is really important now.

We must ask, do any forms of life on earth other than human beings have a capacity for reverence, spirituality, compassion, and a Kantian type of morality? Do they attend church or decide not to do so knowing what it entails, write poetry or science fiction? Their imaginations are limited. Apes have made rhythmic paintings using lines and brush marks, but can not draw objects or landscapes. The argument that animals have consciousness and self-awareness does not make them human-like except in those qualities. They still lack the capacity for visualizing abstract concepts such as deity, morality, reality, justice, quantum mechanics, or psychoanalysis.

Humans' uniqueness prompted Michael C. Corballis, a New Zealand physiologist and psychologist to state in his book *The Lopsided Ape— Evolution of the Generative Mind* (1993): "If human uniqueness is not found in consciousness or awareness of self, I have argued, nevertheless, that there is a fundamental discontinuity between ourselves and other species." The important word in this quotation is "discontinuity." It points out that our difference from animals is a qualitative one, not merely one that is quantitative. If this is true, many sociobiologists' theories of human essence would be false.

Sara Hardy is quoted in *Vignettes: Pitfalls of Evolution* (1994): "The whole message of sociobiology is oriented towards the success of the individual. It's Machiavellian, and unless a student has a moral framework already in place, we could be creating social monsters by teaching this. It fits in very nicely with . . . 'me first' ethos." Morality viewed solely as derived from our past animal behavior deprives it of its meaning. The concept of animal morality described by sociobiologists is incomplete. It fails to reflect voluntary commitments to moral behavior. The "voluntary" marks the point of our discontinuity from them.

Neal Miller, a neuroscientist who gave a major address at the Annual Convention of the American Psychological Association in 1994, began his talk with the astounding statement, "The human brain is the most marvelous thing in the entire universe." He went on to explain that our brain extends beyond instinct and learning to consciousness and creativity. He refers to our "highest mental processes" in nature that enable us to discover a black hole fifty million light years away, "with a mass of a billion times that of the sun." But the human story does not end there. The plot of our story thickens the minute we recognize the fact that our unique human characteristics also impose on our species a unique responsibility. Only if this responsibility accompanies biological research can we hope that genetic engineering will not be abused as its applications expand in the 21st century.

Before this chapter ends, we must ask, "Why do we have a need to put ourselves on the level with animals or see ourselves as akin to computers? What keeps us from accepting ourselves as different from both?" Self-acceptance is as important to a species as it is to an individual. Lack of self-acceptance causes people to reject each others and take refuge in rationalizations. Some animals meet all of their needs—better, perhaps, than we do ours. We differ in that we continually invent new needs for ourselves and always seem to go beyond them once we have satisfied them. In this way, we create our own evolutionary pressures. In the past, people living close to nature went beyond their needs in their imaginations and their fantasies are reflected in their legends and taboos. Technological societies go beyond their immediate needs in a different way—just read some of the advertisements in mail order catalogs!

Could it be that we see ourselves as only a little more than animals because to do otherwise would create guilt? Guilt, indeed, is a widespread human malaise reflected in our various religions and also in the growing number of world-wide cults. Could our guilt come from the suspicion that we are shirking our responsibility as humans? If so, it is convenient for us to picture ourselves as just a little higher than the animals, instead of what the Bible says about us—that we are only a little lower than the angels.

This brings up another matter that we must look into to gain greater self-understanding. Some mathematicians view human life as an embodiment of the information theory. Frank Tipler, a professor and noted author, believes that the human essence could be fully explained as "an entity which codes information" and he states that this process accounts for both the "human mind" and the "human soul."

Certainly, there are animal qualities in us that often take over and dominate our lives. Natural selection does, indeed, play a significant role in human interactions. However, I claim there are more—many more—factors. We have a potentiality to go beyond natural selection, even if some of us do not do so. I differ from the new Darwinists, not so much in how we presently act, but in how we could act if we developed a new mind-set. I believe that the Bible's description of us "a little lower than the angels" (in potentiality) is more accurate than the modern Darwinian view. Low opinions of ourselves as creatures dominated by selfish genes contribute to our loss of self-respect as human beings, and we tend to act accordingly.

I am inclined to agree with Lisa Beachrist, who observed in *Science News* (1995): "The speed and precision of the brain leads some people to refer to it as the ultimate computer . . . the brain is very different. It is able to adjust its own circuitry." Even this is not the whole story. We now have advanced computers that are self-correcting and use artificial intelligence to adjust their circuitry. However, only humans can create computers. Computers cannot create humans except some day, perhaps, they can manufacture ex-humans called *Homo geneticus*. God (or however one accounts for creation) can not be viewed as a computer, although God and humans may exhibit certain attributes that are computer-like. The human essence consists of more than any "complex information processor," even if we define such sophisticated systems as Tipler does—"warm, enjoying, reflecting, worshipping, loving" information processors.

Life consists of a kind of chaos that non-living matter lacks. This gives life unpredictabilities that lead to its vastly increased adaptability and distinguishes life from robots, machines, complex computers, and everything else non-living. In Chapter Five, I maintained that all events—all happenings—create reality. This implies relationship, but not identity or similarity. It is easy for humans to imagine identity where none exists, because in some ways what our brain does may be indistinguishable from what a computer can do. Our responses to virtual reality may be identical to actual reality. Both carrots and cabbages are similar in that we can eat them, but they are not identical merely because they are both vegetables.

We can understand why many people wish to reject the idea that humans stand alone in the cognitive revolution that accompanied our branching out from the primate tree. There is always something about any innovation that is scary. An example is a freely chosen morality without ulterior motives. Once we grasped that a new responsibility must accompany this innovation, it seemed too much for us to bear. In Chapter Four, I cited Jean-Paul Satre's explanation for our modern anxiety. He claimed that we must view it as, "anguish caused by the almost impossible task of giving meaning to a meaningless world." One of the reasons why the world appears meaningless to us is that we wish to see harmony in nature, but observe largely struggle. The Age of

Enlightenment promised us utopias and we see more strife than peace. We worship an omnipotent, merciful God and see misery, disease and deaths of sinners as well as non-sinners from earthquakes, droughts and floods which we would expect such a God to prevent. Compared to animals, we suffer from self-inflicted doubts and insecurities, many of which stem from our ability to ask "Why?" Animals' concerns center primarily on their food supply, procreation, safety, their hierarchical position in the herd or group and probably about things about which we have no knowledge. As far as we know, they do not aspire to achieve wisdom beyond their needs. Are we better off than they? Lord Byron said in one of his poems, "If ignorance is bliss, 'tis folly to be wise."

Wise people make less mistakes than those who are not. To gain an insight into the nature of life we should be aware of life's paradox. Life exists only because of its ability to make mistakes! *Homo geneticus* wouldn't like this definition since it might exclude him from the definition of life. David Brin, a modern physicist, has given us the insightful definition of life which follows. "Life has only one requirement—a system incapable of perfect replication. This is synonymous with our definition of evolution. Naked molecules replicating themselves imperfectly would be a sign of life."

Contrast this thought-provoking definition of life with the conventional one: life has the ability to absorb substances selectively from the surrounding medium and to excrete the products of metabolism back into that medium. The power of growth, multiplication, self-reproduction, dispersal in space, and responsive reaction is manifested by irritability.

The most distinguishing characteristic of life—the capacity to make mistakes—is omitted in the definition given above. Genes make mistakes in replicating themselves. These lead to mutations. Mutations have caused the extinction of species, hereditary illnesses, and physiological malfunction. They have also created diversifications and innovations that enabled survival of a species under the most difficult environmental conditions. The opportunity given life to make mistakes in replication is spelled out in human life by our capacity for free choice. If we examine the laws of the universe and how they are reflected in the laws of life, we will realize that free choice among humans can only be viewed as a "mistake." The basic difference between animals and humans boils down to the notion that we are self-made mistakes while animals are only innocent mistakes of nature created by mutations.

In an evolutionary sense, our human morality as described in Chapter Six must be one of the greatest error ever made in the history of life, since it serves no utilitarian purpose. It is interesting to note that the French Jesuit priest and paleontologist Pierre Teilhard de Chardin, in his book *The Phenomenon Of Man* (1961), described Christ as a "mutation within spiritual evolution offering humankind a new direction for expressing love and sacrifice." Teilhard knew that a mutation is defined as a mistake of replication.

Morality, justice, humility, even fairness are mistakes no other creatures on earth and no computers are capable of making. We are reminded of the proverb To err is human. When we think of our possible future, we must add: Not to err is *Homo geneticus*. We have the distinction of freely choosing our mistakes after considering the outcomes of alternatives. We glory in our mistakes, revere them and try to spread our mistakes throughout the world as behavioral standards. In this we differ substantially from other species, who are merely determined to spread their genes across the world.

When Sir Isaac Newton reflected on his life's contributions, he likened them to a pebble at the edge of an ocean of truth. We often fail to take sufficiently into account the difference between the pebble and the ocean. The ocean Newton referred to can be seen as stretching as much as 100 billion years or more into the future, to the end of the universe—if it has one. Seldom do we consider the possibility that humans may sail that ocean.

As we pursue our research in genetic engineering, it will be necessary to look ahead and redefine our present lives with our future in mind. It has occurred to me that we may not do so because we humans, as a species, are still very young children. Actually, we are mere infants in geological time or, at most, toddlers in the biological time span of life of earth. The problem is that we are extraordinarily precocious children who are apt to overestimate our abilities. Therein lies the danger inherent in genetically engineering ourselves. We have an overwhelming urge to be grown-up, when, in fact, we are small children in a world where animals are the teenagers and adults. This analogy can be appreciated only if we reduce the length that all life has existed on earth to the hours of a day marked on a 12 hour dial of a clock. In that case invertebrates entered the scene at 2:30 P.M. Fish appeared a half hour later at 3:00, amphibians arrived at 4:45 in the afternoon, reptiles at 6:00, mammals at 9:00 in the evening on this dial.

Humans, unprotected and lacking built-in bodily advantages like claws, sharp canines or wooly skins, cautiously entered life's scene only a scant 2 minutes before midnight. Our lack of adequate equipment for survival in a rough world forced us to actualize more impressive potentialities of the universe than other species. Once started, we continued do so. For better or for worse, this may account for human/animal differences most of all.

The more recent fossil evidence that life exploded into new directions in the Cambrian Period, 570–500 million years ago, might upset this timetable somewhat. Nevertheless, humans arrived very late in terms of life on earth as a whole. In regard to the evolution of life, humans are young children. As young children sometimes like to do, we parade around in adult cloths and pretend that we are all grown-up. This might get us into trouble.

To our biological disadvantages we must add that human neonates at birth are the most helpless creatures on earth. Let us include this picture of ourselves as we compare ourselves to animals. It may help us explain a number of things about ourselves not described in our history books. Our journey into the 21st century, armed with gene technology, may be precipitous in view of our immaturity as a species. A mix of childhood helplessness and precocity might be dangerous when making major decisions about our use of gene technology without the firm guidelines that only an incorruptible morality can provide.

Guidelines must identify priorities. The next chapter shall discuss what our priorities might be as we forge ahead with genetic engineering and confront the fact that there is no turning back.

CHAPTER EIGHT

A SEARCH FOR PRIORITIES

Lois Wingerson combined a career in medical research and writing. In her book, *Mapping Our Genes* (1990), she states: "Like Columbus leaving the coast of Spain, the explorers of the human genome cannot see beyond the horizon. They can only imagine what lies there and dream about it. The journey is a little frightening but also very exhilarating. They have persuaded us to pay for it. There is no going back now."

As we stand at the threshold of a new Age of the Genome, we agree that there is no going back now. Therefore, we must prepare ourselves to go forward and try to anticipate the consequences. Beyond the horizon of the researchers lie the applications of their discoveries and the doubt-provoking questions that the freedom of thought produces. The dream to which Wingerson referred could be one of a humankind united in the task of developing an international community of mutually accepting individuals cooperating to employ gene technology for the benefit of all. It could also be a nightmare, where contention over who owns the human genome becomes the dominant issue of the next century. If we wish this dream to have a happy ending, we should begin now to search for a starting point for our hopes and plans to improve the human condition through the use of gene technology.

In the Middle Ages, most people in the Western world were certain that our earth occupied the center of the universe and that a stern, punitive but, if properly approached, forgiving God had created it and ruled over it. As I pointed out previously, people living in those times had no doubts about who they were, why they were on earth, and where they might end up after their death. Their dogmatic religious beliefs and prescribed life-styles were adaptations to the uncertainties and difficulties that touch every human life. Doubt-provoking questions were expediently handled by religious explanations and by blaming Satan. Thus, for the majority of people living in the Middle Ages, all things were accounted for. If these conditions prevailed today, questions of priorities in gene technology could swiftly be answered by the Church and the answers would be accepted as final without question.

We have seen that in the Age of Reason metaphysical explanations and superstitions were replaced by rationalism, science, and freedom of thought. While much was gained for humankind by the "liberation of the spirit," the previous adaptations to life's tribulations no longer served their purposes. Nor has rationalism, as we currently practice it, solved the problems of our times. Molecular biology has raised additional doubts with the discovery of the genetic make-up of life. We must try to resolve these doubts as best we can as we begin to extend the application of gene technology into the next century. Ranking high on our list of priorities for the Age of the Genome we are about to enter is that an effort must be made to regain a sense of community, in which people feel mutual responsibility towards each other without resorting to the paternalism, dogmatism, fears and superstitions of the Middle Ages.

Because the human genome is the community property of our species, it is more than a little disconcerting to have its application subject to the judgment of those with whom we may not agree. This gives us ample reasons for concern, as Wingerson suggests. She points out that we pay for our knowledge of genetic engineering not only with our money but also by permitting research in genetic engineering to take place. The time has come to take responsibility and plan ahead. These plans should not be made by scientists and bureaucrats alone, but should involve many others from as wide a spectrum of the population as possible.

We harbor no illusions that the Big Picture view of gene technology will gain acceptance easily. From Gaia-centered perspectives of Mother Earth, we are less important than the microbes that help maintain the earth's equilibrium. Some have stated that if viruses were to destroy us, the earth itself would not suffer from our extinction. We should note that there are some environmentalists who cast humans in the role of viruses that infect the earth and sicken it. They believe that Mother Earth would divest herself of a burden if our species became extinct. People with these views obviously will have different priorities than those I propose. I believe that our extinction would deprive both the universe and our earth of our contributions that help create their reality, write their histories, and actualize their potentialities. Environmentalists have made it clear that in planning humankind's future we cannot take our earth for granted. I think that neither should we take for granted the contributions made by human existence. Returning once more to the discussion of the previous chapter we can ask: "What is nature's greatest wonder?" Not viruses, not microbes, not fish, fowl, lions or elephants. People attempting to answer this question are nature's greatest wonder.

Most creatures other than humans do not despoil the environment as we do but they cannot serve as role models for us, because they do not preserve the environment on a cognitive level. Geneticists must heed environmentalists, since the extraordinary physical and biological characteristics of our earth give life on its surface a fragility which must play a role in considering priorities for the Age of the Genome. An earth as a mother to only viruses, bacteria, worms, and insects would be deprived of much of its significance. Only humans can perceive the earth's role within the larger universe and relate to it with greater understanding and appreciation than other living creatures.

The religious philosophy of Taoism correctly describes us as co-creators of the universe. Niels Bohr, the Danish physicist who was awarded the Nobel Prize for his work on atomic structure, saw us as playing an additional role. He believed that a description of a phenomenon must include the measuring system as part of the phenomenon. This means that in selecting priorities for actualizing potentialities of the universe we also must become the measuring system that evaluates them. Some would describe our measuring capacity as resulting from our neuroanatomy and psychoneurology. I prefer to see it as stemming from our religions, philosophies, sciences, superstitions, and even our delusions of grandeur. Some people may regret the last two items on this list, but all of them have redeeming features in that they prod us to overcome and circumvent obstacles that could impede the actualizing of our potentialities. Our limitations stem from the fact that we must rely on a Big Picture view as perceived from our point of observation. Recognition of the limits of our perception can help us decide where to begin our search for priorities.

I propose that high on any list we should place our desire that human beings will continue to survive. My reason for this differs from those given by sociobiologists, who tend to view all living things on the genetic level. The Big Picture view suggests that we have hardly begun to mine the treasury of unactualized potentialities of the universe. For this reason, we remain unfulfilled as a species that is gifted with the potential to do more. Let's also admit that aside from these considerations it's pleasant to feel that in spite of our shortcomings we will be around for quite a while longer, especially if eventually we will be able to enjoy a happy human togetherness. Therefore, it should come as no surprise that I propose human survival as our first priority.

The benefits of genetically engineered medical products are more obvious to the public and therefore gene technology for medical use is generally readily accepted. Biotechnology can produce insulin and the cancer-fighting drug interferon alpha. There is even acceptance of the ongoing research producing human hemoglobin and other medical products utilizing pig's blood if it will enable victims of diseases to live longer, healthier lives and help humankind to survive.

In order to survive, we must maintain our can-do spirit. Our ability to use gene technology could not have come at a more propitious time. I have previously called attention to our competitors, the tiny microscopic genetic engineers that could bring about our extinction. Common bacteria that cause human illnesses as certain ear infections and pneumonias are now becoming increasingly resistant to treatment and may in the foreseeable future become untreatable by presently known medications. Certain strains of bacteria that cause wound and blood infections no longer respond to the antibiotics available today. Some structures on bacteria called plasmids carry genes that enable them to resist antibiotics. As I stated previously, they can share these plasmids with bacteria of different strains, enabling the resistance to antibiotics to spread rapidly. The Age of the Genome may gain the distinction of ushering in a post-antibiotic era.

Those who consider our contributions to the history of the universe as important and give our survival the highest priority are concerned about the loss of the effectiveness of current antibiotics. Referring to the new immunity of bacteria to our antibiotic arsenal, Dr. Mitchell Cohen at the Center for Disease Control and Prevention was quoted in the *San Diego Union-Tribune* on February 20, 1994, as stating, "Some common bacteria evolve into wholly untreatable strains. It's potentially an extremely serious problem." He estimated that new drugs necessary to treat diseases resistant to current antibiotics are at least five or more years away. In terms of our priorities, we must attempt to design our genetic engineering to be superior to that of the tiniest microorganisms. Therefore, as we wage our war on inherited diseases, we must expand our efforts to include the use of gene technology as a weapon against some of our smallest and deadliest foes. I propose an expanded war against all human diseases as a high priority. Would this also include mental illness? There is impressive evidence to support the existence of genetic factors in certain types of mental illnesses.

I do not think we know enough at this time about how genetic factors contribute to mental illnesses to make decisions on this subject now. It would be wise to view genetic engineering as a last resort in those cases where it might deplete the can-do spirit that helps make us uniquely human. Notwithstanding, if at a later date we gain the skill to alter the genetic factors that lead unequivocally to mental illnesses, we would rid humankind of the miseries and anguish that accompany these conditions. In the future, we may have to weigh the question of lifetime imprisonment and the death penalty against redesigning antisocial individuals who are chronically unable to resist committing crimes. There may be considerable danger of overplaying our hand in such an effort; if so, we must balance this danger with a responsibility to alleviate the suffering of victims of remedial genetic abnormalities of all types. Another consideration in setting priorities takes originality and creativeness into account. Since the dynamic push of the universe leads towards active happenings, it is reasonable to give them a higher priority than passive ones. I do not view relaxing, meditating or living a quiet, thoughtful life as inactivity. Instead, I think of inactivity in the sense of not entering into the flow of the dynamics that gives us our unique spirit. I would rank thinking over not thinking, working over not working, creating over not creating. I would place originality that brings novelty to the world high on our list of priorities. Each new happening expands reality within the universe, just as the sum of our own external and internal experiences contribute to the reality of our own lives.

There is a special charisma in the phrase "for the first time in history," whether it pertains to the beginning of our own lives as celebrated by birthdays or to other new events that take place in our lifetime. For example, an exact copy of a masterpiece of art has far less value than the original painting even if no one can distinguish it from the original. Anything that earns the label original creates a special kind of excitement. An event recurring later in the identical or even in an altered form only activates the potentiality of repetition. We see this in records set in sporting events, the discovery of Antarctica, a human footstep on the moon, a new species discovered, or some cosmic event witnessed for the first time. Today's aircraft fly faster than the speed of sound and rise high above the earth's atmosphere. Yet the monument dedicated to flight maintained by the National Park Service was not erected to honor today's pilots. It honors Orville and Wilbur Wright, who made the first engine-powered flight although it only lasted 12 seconds in a plane that was able to rise only inches above the ground. It was a first, and for that reason its significance outweighs all of the flights that followed. The thrill that we get at the idea of being the first to accomplish a task reflects our role as co-creators in a universe that expands with every first that occurs. Of course, not every time a new event takes place is it favorable to us. Nevertheless, the idea of "for the first time in history" inspires awe and, if it is beneficial to us, causes jubilation. The excitement when a new species is discovered, even if it is of no consequence to us, stems in part from the fact that it reveals to us another actualized potentiality of the universe.

Taking the subject of "for the first time" to geology, we are reminded that climatologists have found that the earth has been subject to violent climate swings ranging from ice ages to extreme heat. In the last 250,000 years of climate history, the most recent 100,000 seems to have been the only period in which climate remained relatively stable. We were able to build our civilizations during what may have been the only time when climate was sufficiently favorable to allow the beginning of agriculture, the development of culture, and the evolvement of an industrial society. It is predicted by some concerned environmentalists that the benign stretch of stable climate upon which we depend to maintain our quality of life may soon end. Between 2200 and 1900 B.C. thriving civilizations along the fertile valleys of the Tigris and Euphrates rivers, in what now is Syria and Iraq, suffered an abrupt climate change which brought about 300 years of drought and severe dust storms that have been linked to the collapse of that "cradle of civilization."

Nature alone caused this devastation. Environmentalists warn us that today we face a similar threat to our civilizations, but this time the destruction would be our own making. Our age has contributed a new page to the long history of challenges that we have confronted. These challenges involve undoing the damage we have inflicted on our environment through the careless use of our natural resources and the pollution of our air and water that are the results of our zeal to be candoers, regardless of consequences. To some extent, this trend can be reversed by the use of biotechnology, and I consider further application of bioengineering a high priority.

In the Age of the Genome, industrial chemicals may be grown by bacterial colonies. Some day it will be possible to replace mechanical engineering with bio-engineering. Molecular scientists using genetic engineering may, in the foreseeable future, develop biological sources of reusable energy that could replace fossil fuels. They envision a time when non-polluting power will be extracted from living microorganisms or their sub-molecular parts that would reduce entropy and, at the same time, produce non-polluting energy with considerably less effort and expense than is possible today. One of our greatest human needs, fresh water for a thirsty earth, will be obtained in the 21st century by salt-devouring microorganisms. Osmosis and evaporation are not convenient methods to assure us of an adequate, reasonably inexpensive water supply. Already, we have subjected bacteria to nutrients laced with arsenic and after numerous mutations in that poisonous milieu bacteria thrived on it.

At some future date, microbes may consume our ever-increasing waste while producing methane gas that could fuel automobiles and locomotives. This use of genetic engineering could bring with it a new industrial revolution fueled by bioenergy. In essence, a technology that uses genetically engineered microorganisms to perform labor represents a breakthrough similar to the one that occurred when domesticated animals were first utilized for transportation and agriculture approximately ten thousand years ago. This breakthrough changed human life then and increasingly threatened the ecology by overgrazing and depletion of the fertility of the soil. The wide-spread use of bioengineering may again change human life, this time saving our planet from further destruction. Moffat (1994) reports that within the past few years the mining industry has turned with increasing frequency to biomining that avoids crushing the earth, creating extreme heat as well as toxic chemicals. This method uses bacteria that separate metals from ores. As more high-grade ore is depleted, biotechnology permits the economical extraction of minerals from low-grade ores.

Agriculturalists have manipulated nature for years utilizing cross-breeding to develop improved varieties of such crops as corn and potatoes. With little opposition, they have crossed tangerines with grapefruit to produce tangelos. Genetic technology goes beyond this to give wheat, rice, and other grains the ability to resist droughts and diseases. Tomatoes, for example, have been genetically altered to ripen longer on the vine while remaining firm for picking and shipping, making them more flavorful and giving them a longer shelf-life at the supermarket. Fruit and vegetables can be given year-around availability. Such breakthroughs could be hailed as a solution to the world's impending food shortages.

In setting priorities, we must take into account what people find acceptable. For example, more than half the people who participated in a survey on public attitudes conducted by the U.S. Department of Agriculture at North Carolina State University found it repulsive to use animal genes to help increase the value of food products for human consumption. A bioengineered gene recently designed to help maintain a desirable texture in frozen strawberries serves as an example.

Molecular biologists are able to isolate a gene in the DNA of an arctic flounder which prevents the fish from freezing in icy waters. Geneticists using today's technology extracted this gene from the flounder and reproduced it in a bioengineering laboratory. They then inserted the gene into the DNA of a strawberry and this delicious contribution to our gustatory joys was altered to retain its texture and taste when frozen and stored. Half the respondents to questionnaires used in the survey admitted they feared toxic and allergic reactions to such bioengineered strawberries. Are these fears justified by the possibility of unanticipated dangers in bioengineered foods? Or will such concerns, in time, join the long list of baseless objections to innovations?

The agricultural biotech industry is expected to develop into a 50 billion dollar industry by the year 2000. Some scientists do not rule out the possibility that bioengineered foods might inadvertently activate a toxic gene in the DNA of a newly created genetically-altered food. Carefully testing the safety of biologically engineered foods before they become available to consumers ranks high on our list of priorities. Still another valid concern relates to the danger of upsetting the delicate balance of nature when we introduce new species into an environment that is not their natural habitat. The accidental escape of the destructive gypsy moth from a scientist's lab damaged our ecology, and the inadvertent import of zebra mussels to the U.S.A. aboard a foreign ship continues to cause serious environmental damage to the waters of our Great Lakes. One may ask, "Do we really know what problems we may create when we genetically alter plants and animals?" Careful investigation of possible negative effects of all biogenetically altered forms of life ranks high on our list of priorities.

When a food is confirmed as safe and nutritious, objections may subside but distrust at the idea of altering nature with biotechnology may still linger. Over half of the respondents to the previously mentioned survey at the University of North Carolina were opposed to the use of gene technology when it involved the interchange of animal genes, not only because the idea was disturbing to them, but more particularly because they felt it was morally wrong. As seen by the results of the opinion poll, the perception of morality in the Age of the Genome will have a major impact on how gene technology will be put to use in the 21st century.

We have said much about morality so far. It is now time for us to ask, "Can morality survive in a world that requires struggle and ongoing competition?" It is important that, at this point in the book, we now explore the roles of harmony and struggle in our lives. It is a dimension of evolution that requires further inquiry. If we do not give sufficient thought to whether human life thrives best on harmony or struggle our picture of the future of gene technology will be incomplete.

CHAPTER NINE

OUR ROLE IN NATURE

Both ancient and modern literature contain conflicting views of the roles of human beings in the world. One of these portrays harmony in nature and a desire for it among humankind. The other describes strife and struggle as natural and inevitable. This view glorifies wars and conquests.

We must search for our role between these extremes and also remain aware that the need for both harmony and strife may exist within human nature. We shall examine each of these qualities to help us avoid errors when we are able to apply gene technology both beneficially and destructively. The outcome of a search for our role in nature depends to a large extent on our perception of harmony and struggle. Our conclusion about them will serve as the foundation upon which we will build our future world.

Harmony, as we shall use the word in this chapter, signifies an equilibrium that leads to the perception of an underlying serenity in the universe. A key word to remember when thinking of harmony is balance. Imbalance always requires an effort and sometimes a struggle to restore equilibrium. We use the term struggle to mean imbalance and confrontation with difficulties in life. *Webster's New World Dictionary*, illustrates the use of to struggle: "to make one's way with difficulty, as a struggle through a thicket." A view representing harmony, on the other hand, sees the same journey as one largely consisting of cooperation and symbiosis.

The question of which of these contributes most to human nature must take us beyond individual opinions and culturally-shaped attitudes. In order to examine harmony and struggle meaningfully, we must retrace our steps to the prehistory of modern humans. Beyond this, we must see how conflicting concepts of evolution and the latest discoveries in cosmology play a part helping us locate our role in nature. In this search. we must not omit the views handed down to us by various religions.

Anthropologists have found that an early hominid (human-like) species named *Homo habilis* was among the early progenitors of modern humans. Over the period of approximately two million years that *Homo habilis* evolved into the modern human being, a gradual change took place in how humans perceived the world. Instead of merely observing that changes occurred in their environment, humans began to ask themselves how and why such changes occur. Early in prehistory, hunters and gatherers must have wondered why the sun and moon didn't fall to earth and what caused the regular cycles of day and night. They watched stones tossed up return unaided to earth. They felt the wind on their bodies and saw it made bushes and tree branches sway. They observed rain fall from the sky and ocean waves roll rhythmically. They may have felt the regular beat of their hearts and realized that there was rhythm within themselves as well as in nature.

Our ancestors observed that plants grew out of the earth and realized that all living things begin with birth. They knew that they shared birth and death with animals. As some people in isolated parts of the world still do today, they must have reasoned that a variety of spirits inhabited all things—trees, stones, animals, and people. They believed that these spirits could cause things to happen. Our ancestors were aware of the fact that everywhere predators survived by killing weaker animals for food and that struggle existed in nature, as well as the harmony they saw in the sky and on earth. When early humans witnessed catastrophes such as earthquakes, droughts, floods, and forest fires, they were reminded that nature could be terrifying as well as nurturing.

Thoughts of a spirit world must have occurred to the Neanderthal people, who as long as 40,000 years ago buried their dead in shallow graves with their flint tools beside them. Lifelike paintings and drawings of woolly mammoths, bison, wild horses, and reindeer still can be found on many cave walls in Spain and France. We may speculate that these sites served as centers of religious rituals. The artists were Neolithic humans who chipped well-designed stone tools 20,000 to 30,000 or more years ago. They too must have believed spirits inhabited the animals they portrayed. They could observe that animals as well as people exhibited affection, courage, shrewdness, and fear. All natural phenomena, even rocks and wood, were thought of as inhabited by spirits, a belief known as animism.

About ten thousand years ago, agriculture gradually replaced foraging and people were able to accumulate surplus food that could be stored for future use. This provided them with increased leisure to try to understand what caused things in their environment to occur. They believed that supernatural beings, usually gods or animals with human-like qualities, were responsible for all events that happened. Not all gods lived in harmony with each other. As with humans, strife and jealousy existed among them. Once humans had time to think about the world, it seemed much more complicated than it had appeared to our earlier ancestors, who were too busy trying to survive and perhaps not yet capable of thinking about abstractions.

In 500 B.C., the ancient Greek philosophers theorized about the presence of harmony in the universe. "Cosmos" comes from the Greek word *kosmos* meaning order and harmony. Philosophers acknowledged that strife existed in nature but when they looked at the starry skies, which they likened to a beautiful ornament, they found harmony more fitting their description of the universe than struggle. According to the Old Testament, the first humans were expelled from the Garden of Eden, which was pictured as a paradise where harmony prevailed. Harmony, it was predicted, would return to Earth with the coming of a Messiah. Buddhists view harmony as inherent in an all encompassing concept of an Ultimate Reality. Struggle becomes a counter force to harmony and results from human ambitions and the pursuit of pleasures. Hinduism describes harmony as the balance of three tendencies: action, stability, and decline. Taoism finds harmony in the interplay of opposites such as in Yang and Yin as reflected in earth/sky, day/night, male/female. Mohammed's teachings emphasize brotherhood, and the Koran offers directions for living that would replace strife with harmony. It is clear that the modern world religions reflect the harmony for which people yearn while, at the same time, we find them always struggling.

Eastern views of harmony do not contradict the can-do perception of humankind. Religious teachings that lead to harmony through God do not deny that struggle may occur in the world. Considerable space in the Bible is devoted to strife, as well as to people's inner struggles. All explanations of nature's way must take into account the dynamism of the universe, which Taoism describes as a "creative force in nature that depends on humans as co-creators." What might Taoists reply if asked, "Since we are co-creators, isn't genetic engineering a legitimate pursuit for humans?" They might reply that the things nature creates are not appropriate for humans to create. They would probably consider that designing life would be best left to nature.

In the 18th century, some naturalists in France, England, and the United States saw the evidence of harmony in the way nature controls over-population. They observed that when one species becomes too numer-ous to insure its survival, nature restores balance with natural disasters and by increasing the number of its predators. This led people to view nature as an all-wise guiding force. But others observed that many animals have to defend their territories. Males of some species engage in a kind of psychological warfare when competing for mates. The food chain that sustains life is by and large one of cruelty and death. This is expressed in the following quotation of unknown origin. "Every morn-ing in Africa a gazelle wakes up. It knows that it must run faster than the fastest lion or it will be killed. It knows that it must outrun the slow-est gazelle or it will starve to death. It doesn't matter if you are a lion or a gazelle. When the sun comes up you had better be running."

The "running" means, running away from something to remain alive or toward something to get the food required to stay alive. According to this observation, ongoing struggle for existence seems the most accu-rate interpretation of nature's way. Struggle is seen by many as the process by which species that adapt to their environment survive while those who are unable to do so perish. Charles Darwin published his theory of evolution in *The Origin of Species* (1859), often referred to as "the book that shook the world." It shook the world not only because it seemed to contradict the Bible, but also because it replaced the idea of harmony with one of struggle as the dominant force in nature. Later Darwin wrote *The Descent of Man* (1871), in which he expanded his earlier theories. Darwin considered the survival of the fittest through struggle the most important aspect of natural selection. Survival was a species' reward for competing successfully with other species for the same limited ecological niche. The theory of evolution profoundly re-vised people's view of our role in nature. In Chapter Eleven, we shall examine Darwin's theory and find it incomplete. Without involving the-ology, we shall consider it as only a fragment of reality and therefore misleading when used to creating a mind-set for the application of gene technology.

In the aftermath of Darwin's theory, Herbert Spencer, the influential English biologist, philosopher, and writer asserted that the elimination of the weak of a species by natural selection through competition would eventually lead to a state of equilibrium and harmony. He equated successful adaptation with ethics and saw the failure to adapt as a moral evil. In essence, Spencer believed that nature attempts to rid itself of its "mistakes." This included eliminating people who are unable to compete. Spencer viewed any nation that helped to keep the unfit alive as doomed to extinction. He championed *laissez faire*, the idea that a state should not interfere with private initiative in any manner. Consequently, Spencer opposed state banking, state-supported education, state charity for the poor, and state support and regulation of housing. He believed even taxes and government postal services were undesirable.

One can't help wonder what role genetic engineering would have played in Spencer's philosophy of noninterference with natural selection, had he been able to anticipate that it would be discovered in the future. Would he have placed genetic engineering into the same undesirable category as taxes and postal services, or would he have had to revise his view of non-intervention if geneticists demonstrated that gene technology could some day redesign people to make them more fit? Spencer's antigovernment ideas are worth reviewing, since they are being revived by modern movements that have sprung up throughout the world, especially in the United States.

After Darwin's theory of evolution had gained world-wide attention, some sociologists attempted to introduce the biological principles found in evolution into social reforms. They warned their government leaders not to interfere with the use of competition to eliminate the unfit since, they felt, competition was nature's way of improving life. These proponents of unfettered competition were called Social Darwinists. However, other writers on social issues strongly opposed their views. One of these was the influential Scottish preacher and writer, Henry Drummond, who delivered a series of lectures in 1894 at Harvard University entitled *The Ascent of Man*. He used ascent to emphasize his opposition to Darwin's perception of human's descent from other forms of life. Drummond didn't deny that there was struggle for survival, but viewed it as only a first step in what he called, "the struggle for the lives of others."

The impact of Drummond's insight has been generally lost. His perception of struggle too often fails to enter into consideration in discussions on whether harmony or struggle represent nature's way. With our large variety of options, nothing in nature suggests that struggle in human life must necessarily be against something or someone. A struggle for something does not diminish its challenge. Struggle for something is often called striving. Struggle, striving, harmony, balance—are more than questions of semantics. Both struggle and striving represent efforts that could lead to making creative changes which actualize new human potentialities.

Edward Bellamy, an American novelist, went one step further than Drummond when he claimed that many who succeeded in the competitive world of the Social Darwinists were really the unfit in terms of what counted most—humanitarianism and compassion. The debate continued as one of Bellamy's contemporaries, Kenneth Walker, a prominent British surgeon and scientist, declared Bellamy's views "shallow." In his opinion, humankind had risen from stage to stage mainly by means of competition and ongoing struggle.

Jean Jacques Rousseau, the political theorist of the 18th century period of Enlightenment, extolled the harmonious life of people living undisturbed in natural surroundings. The 20th century Nobel laureate Francis Crick, co-discoverer of the structure of DNA, observed that there was strong selective pressure for cooperation within small groups of people while at the same time there was a concurrent hostility towards neighboring, competing tribes. He points out in *The Astonishing Hypothesis* (1994) that "Even in this century, in the forests of the Amazon the major cause of death among competing tribes in remote parts of Ecuador is from spear wounds inflicted by members of rival tribes." One reason we find perceptions of harmony versus struggle confusing is caused by the fact that they coexist.

Edward Harrison, a professor of physics and astronomy, wrote in *Masks of The Universe* (1985): "Some people have the impression that the physical universe is a world of extreme violence, explosions, and cataclysms . . . the closer we examine the design of the universe, the more we marvel at its harmony and beauty." Einstein proposed similar views. It is not difficult to see how the fixed laws of the universe give cosmologists the impression of order and elegance. Yet healthy can-do people upset balance every time they change the status quo. Ernst Mayr,

Professor of Zoology at Harvard, found that "the concept of the benign balance of nature" could not be sustained. He pointed out in *The Growth of Biological Thought* (1982) that the idea of sustained harmony in the world became untenable when fossil records revealed that many species had become extinct and when geologists learned how greatly the world had changed geologically throughout the ages.

We need not consider the issue of harmony versus struggle from the view of geological ages and extinct species. It can also impact a person in the span of his or her own lifetime. In a questionnaire that asked older people what they would do differently had they a chance to live life over, an 84-year-old woman wrote: "I'd like to make more mistakes the next time." The mistakes she referred to had nothing directly to do with struggle or harmony yet, they had implications for both. The woman felt that she had played her cards in life too cautiously. She explained that she had tried hard never to upset anyone or anything. She was bent on maintaining harmony and balance in her relationships with people and her environment whenever she could. In retrospect, she thought that greater accomplishments associated with struggle and controversy might have enhanced her life.

We know that the balance that leads to harmony exists in the universe. However, if balance were the primary force and people would adapt themselves to it, we would have to become accustomed to watching sporting events in which every team's score tied every other team's score all season long. Had harmony existed in the days of chivalry, no medieval knight would have wanted to unhorse another knight in an attempt to gain his special lady's admiration. Adolf Gruenbaum, writing in *The World of Physics* (1987) observed: "The universe around us exhibits striking disequilibrium of temperature and other inhomogeneities." We can be glad it does, because without the imbalances and "inhomogeneities" many of the universe's exciting potentialities would have remained forever dormant and our species, *Homo sapiens*, would have been among them.

"Two contrary laws seem to wrestle with one another," said Louis Pasteur, the keen observer of nature who first proposed the germ theory of infection. "The law of blood and death always anticipates new methods of destruction . . . the other law, of peace, work and health, always anticipates new means for delivering man from the scourges which beset him." Most people are aware of this. We observe this contest almost everywhere in nature. The HIV viruses, after becoming active in the human body, kill more than a billion white cells a day. With genetic engineering, we are making some headway in our battle with this persistent killer. The psychiatrist, Karl Menninger, who we quoted in Chapter One, offers the following thought on the pervasive role of strife, " . . . it is easy to see that all work represents a fight against something. The farmer plows the earth, he harrows it, pulverizes it, he pulls out weeds, or cuts them or burns them; he poisons insects and fights against droughts and floods. . . ."

Menninger points out that the "destructiveness" of the whaler is different from that of the lumberman, the miner and the surgeon, but notes that "all of them are working against something to master a situation or a material to produce something else in the end." Menninger describes a world in which struggle (I prefer this term to "destructiveness") offers us the chief opportunity to use our human talents. Struggle seems to rule our lives.

Nevertheless, the concept that harmony exists throughout the universe is easier to accept than one of ongoing struggle. We experience a sense of harmony with nature when we are in close contact with it. There is something vastly satisfying about the idea of harmony that seems to come from a source deep within us. We enjoy the beauty of the sky and our life-sustaining earth with its many hues and forms. We sing and dance to the rhythms of the universe as they pulsate through us. Often at such moments we loosen up and experience harmony as we feel as one with all else that exists in the cosmos.

But how appropriate is it at those times to consider the universe peaceful, when nearby a snake swallows a rat, a hawk zooms down to seize a rabbit, and a spider devours a fly trapped in its web? Nevertheless, it is more pleasant to think of song birds as singing from *joi de vivre* than as a warning to other birds, "This is my territory. Stay out!" We can admire a peaceful scene of grazing gazelles, but an ethologist would know that the peace was attained only after vigorous horn butting to establish the dominance of the most aggressive male. Other eligible males excluded from the jealously guarded harem eagerly wait for an opportunity to take over the role of the dominant male.

The microscope reveals the Big Picture view as well as the telescope does. The Big Picture view of the universe encompasses the small picture view of viruses and other microbes that infiltrate our cells and subvert our DNA to make copies of themselves at our expense. The T-cells of our immune system that protect us against our ever-present miniature enemies belong to the Big Picture view of the dynamism of the universe acting through life. Wherever we look, struggle swirls around and within us, but since we prefer harmony, we would rather not think of it. The notion of peaceful harmony as characteristic of the world is a human invention.

The idea of disequilibrium is unappealing. We find a picture hanging at an angle on a wall disconcerting. Even in someone else's home we feel an urge to straighten it. We want to see things balanced. Balance is a potentiality of the universe we seem inclined to actualize. Yet it can be argued that we need imbalance to give us our can-do spirit, just as the Big Bang (physicists describe it as a space warp) required disequilibrium to make it bang. Could our wish to straighten a picture that hangs askew reflect an unsuspected meaning in human life? Does it define our role in nature? If so, when the Big Picture hangs straight, we can fantasize that perhaps the work of the universe will have been accomplished. When the final homeostasis occurs many billions of years in the future, most cosmologists predict it will bring all things to a complete and final halt. It is interesting that other scientists go on from there to theorize that another universe may emerge from the demise of our own, while still others speculate that there may not be a demise at all, since the universe could go on expanding forever.

Some cosmologists theorize that when the universe eventually collapses into itself and shrinks into a measureless gravitational singularity sometimes referred to as a "black hole," its irresistible gravitational pull will suck all things into itself—everything, even light. Then, perhaps, the can-do challenges derived from imbalances will fizzle into a final frozen balance at last. The renowned cosmologist Stephen W. Hawking states in *A Brief History of Time* (1988) that after the universe has become a "black hole" it must settle down into a state in which it no longer will be pulsating. After Hawking's book was published, additional theories about black holes were proposed that include gravitational waves surrounding black holes discovered in some regions of our own galaxy. If we accept the view of reality proposed in Chapter Five, no gravitational pull will ever erase the history of our universe's actualized potentialities that include you and me and Stephen Hawking and the black holes themselves. As I proposed in that chapter, all things are forever. It is a needed cheerful thought in what appears as a dismal conjecture of the fate of the universe.

Let's look at the meaning of balance in the field of mental health. The term "mentally unbalanced" is another way of referring to psychosis or mental illness. It is a mistaken notion, since emotionally disturbed people achieve balance by means of their symptoms, while the rest of us continue to struggle with imbalance within and outside of ourselves. Harmony has been defined as agreement and this fits schizophrenics better than other people, since delusions enable the mentally ill to agree with every thought that comes into their minds, no matter how absurd it may seem to us. Psychotic symptoms permit mental patients to deal with conflicts with which their brain, for structural, neurochemical, or intolerable environmental pressures, cannot cope. Psychiatrists have met "generals," "admirals," "billionaires," and occasionally even a "god" in the locked wards of state hospitals.

This is an example of the pervasive thrust inherent in the universe to actualize its potentialities that is manifested in the human can-do spirit. The easiest way to obtain can-do satisfactions is to develop delusions of grandeur. The drive to do so is demonstrated even among those who have lost their judgment and reason. A significant amount of all the crime committed world-wide stems from an otherwise unavailable opportunity for actualizing alternate kinds of potentialities of the universe. Those who are excluded from this opportunity become withdrawn or dangerously antisocial or even mentally deranged. This seems to point to what our role in nature is. Humans, better than any other living creatures on earth, can reach deeply into the contents in the cosmic storehouse to find where some potentialities are hidden. Every time mentally healthy people actualize potentialities of the universe, they create new realities which serve as catalysts for producing others. Thus we play a major role in creating the world in which we live. Depending on how *Homo geneticus* is engineered, he may take pride in his ability to exist without actualizing anything beyond automatically existing. Harmony will then prevail.

Harmony evokes scenes of quiet pastures, lotus flowers, and self-effacement. On the other hand, struggle brings to mind competition for ecological niches, horn-butting, and the mapping of the human genome. Were we living in a world of harmony, would there be longing for another more exciting world? Quiet pastures alone cannot sustain life.

Some people must hunt, fish, or till the land in order to eat. The fear of death may in part consist of a fear of being forced to accept a state of uninterrupted harmony. Perhaps that is one of the reasons normal people fear death—even those who visualize an afterlife in heaven. Of course, they are mistaken in their view of heaven, nevertheless some equate it to the earthly lives of member of the *Homo geneticus* species. Pity them. Hell offers a poor alternative.

Let us look at the question from another point of view. A case for the existence of harmony on earth was made in the Gaia hypothesis formulated by James Lovelock, the English inventor and biochemist in 1970 and Lynn Margulis, professor of microbiology at the University of Massachusetts in 1980. Gaia came of age with the publication of *Scientists On Gaia* (1992), edited by Schneider and Boston. Gaia maintains that symbiosis, the intimate living together of different organisms for their mutual advantage, causes evolution instead of struggle or competition. Since the theory of evolution represents a small picture view of life, it is refreshing that alternative hypotheses exist. According to the theories of Gaia, symbiosis led simple organisms to climb the phylogenetic ladder that eventually brought them to more complex and sophisticated forms of life. However, critics of Gaia charge that it introduces a mystical quality into evolutionary theory that in some ways goes back to the idea that microbes, algae, trees and rocks are inhabited by the spirit of a "mother earth goddess."

Phil Shannon, long active in the environmental movement in Australia, writes in *The Skeptical Inquirer* (1992) that Gaia brought "a needed self-effacement to an environmentally dangerous, technologically powerful, self-centered species." He refers to humans, of course, but where would one find a species that is not self-centered? One could consider environmental concerns themselves as self-centered, inasmuch as we depend on the environment for our enjoyment of life as well as for our survival. Nevertheless, Shannon may have a point. "Self-effacement" and the attitude derived from our self-image as humans will be of primary importance to us in the 21st century. Self-effacement is a behavioral innovation in life, if not actually in the universe as a whole. One had to dig deeply into the welter of the world's potentialities to discover and actualize self-effacement. This is because it conflicts with our view of a Darwinian world order.

As we approach the end of our discussion of our human role in nature, we shall turn the harmony/struggle coin over to its other side. We can look at harmony in a different light if we put it into the category of abstract ideas, together with morality, humility, self-effacement, and other human qualities we value because we want to live in a world in which they exist. In this light, harmony becomes worth seeking even if it takes struggle to achieve it. In that case, we would not have to mourn the disappearance of struggle in our lives, since we will always require it to maintain harmony.

Struggle is an inevitable component of life, but we can use it as Drummond did. We mentioned him earlier in this chapter. Drummond saw value in struggle when it becomes a "struggle for the lives of others." No one represented this point of view more forcefully than the central religious figure in the Christian religion. No one has argued against it as eloquently as the philosopher Nietzsche, author of, *The Antichrist*. Both Jesus and Nietzsche left humankind instructions on how to achieve a new mind-set. In the next chapter, I shall not discuss their teachings as a theologian nor as a philosopher but as a behavioral scientist, my own field of specialization. Which one of their views we adopt as we move on to the 21st century will determine how future genetic engineering will be carried out.

CHAPTER TEN

A COMPARISON OF JESUS' AND NIETZSCHE'S ANSWERS

The similarities and differences between how the 19th century German philosopher, Friedrich Wilhelm Nietzsche and the historical Jesus of Nazareth viewed human life will help us become aware of some of our options in the 21st century. Separated by 2000 years, Nietzsche and Jesus tried, each in their own way, to teach humankind important truths that they believed were hidden from the world. Each presented alternate mind-sets for his time.

By the 21st century, molecular biologists will have succeeded in mapping the position of all of the approximately 100,000 genes on the DNA located in the human genome. At that time we shall be able to prevent many of our inherited diseases and will have gained the power to redesign all forms of life , including our own. It will give us unprecedented power. It could greatly benefit us and enhance our lives, but it could also bring disaster if we yield to the temptation to misuse it.

It is appropriate that we listen to the voices of those who have had a strong impact on our civilization and devoted their lives to guiding us to prevent us from making errors in the way we choose to live. The list of those qualified is a long one. I focus on two persons in our history whose teachings are particularly pertinent to the task of adopting a mind-set for the time when we shall gain a vastly increased power over nature. I do this without any presumptions that I could add anything of value to either theological or philosophical interpretation. My goal is to write as a behavioral scientist without going into the theological aspects and rely on what others have said or written on the subject.

The sayings attributed to Jesus and the writings of Nietzsche both present perceptions of human nature and ways to improve it. I shall briefly review certain aspects of the lives of Jesus and Nietzsche and some of the events that contributed to the formation of their ideas. The task is not a simple one. Approximately 60,000 books were written about Jesus in the 19th century alone and each of them gave his life a somewhat different emphasis. The observations and descriptions in this chapter do not intend to contradict views others may hold. They are gathered from a variety of sources and they fit into the context of this book.

In *Who Was Jesus?* (1970) Colin Cross, a writer of history and biblical archeology, makes my point when he compares the uncertainty of what Jesus actually said with the doubt some scholars have about the authorship of the plays attributed to Shakespeare. Cross concludes: "the works of Shakespeare exist, whoever wrote them, and in the same way the parables and preaching of Jesus exist whatever their precise origin." In regard to historical origins, Oscar Cullmann, professor of Early Christianity at the Sorbonne, in *Jesus and the Revolutionary* (1970) makes the point that historians often face the dilemma that facts and philosophy are interdependent in that facts must be interpreted within the framework of the person's philosophy and religion. Cullman refers to Albert Schweitzer's book *The Quest for the Historical Jesus* (1906) in which Schweitzer cautioned against a modernized, distorted view of Jesus derived from personal philosophies. Reflecting on this same point, I wrote in *Conflict Resolution* (1986, p. 2), "History testifies that conflicts center on different interpretations of what is true."

E. J. Goodspeed, former Chairman of the New Testament Department at the University of Chicago, in *The Life of Jesus* (1950), tried to reconstruct some events in the life of the historical Jesus by using available information and the logical consistency of historical data. I borrow the following description of Jesus' early years from some of Goodspeed's conclusions. Jesus lived in Nazareth and grew up in a large family, in a Jewish home typical of the times. Because of widespread poverty and the occupation of the land by Roman conquerors, there was great discontent throughout Israel. The Jewish religion as practiced then seemed unable to solve the peoples' many problems. This affected the young Jesus who was concerned about their unhappiness.

In his search for answers to the questions that troubled him, Jesus went to the banks of the Jordan where his cousin, John the Baptist, preached to large crowds. John baptized some of the people and Jesus was among them. After he emerged from the river's muddy water, Jesus experienced a transformation. He saw the world and people's problems in a new and more hopeful light. He wanted to share his insight with the people of Israel because he believed that it would enable them to override their seemingly insoluble problems. In effect, it required them to reverse their view of the logical order of things. He decided to dedicate his life to this mission. He wanted to offer his people a new mind-set.

"Call me rabbi," the Hebrew word for teacher, Jesus told those who asked him how they should address him. This gives us a strong hint as to what Jesus may have thought if he were living today regarding genetically implanted intelligence. Teachers who believe that people could benefit from being their students would not advocate the use of gene technology as a substitute for their instruction. Jesus used parables to teach the people that they did not have to despair. He declared that the Kingdom of God had arrived and could turn their lives around. He promised that in the Kingdom of God the last—the downtrodden and suffering—would be the first, and the first—the proud, who thought of themselves as superior—would be the last. If those who had sinned repented, they would be forgiven and could enter the Kingdom of God. According to some Christians, the Kingdom of God has meaning restricted to the phenomenon of Jesus. Others see it as compassion, concern, and love among people.

Nietzsche was born in Prussia and lived from 1844 to 1900. His father, a Lutheran minister, died when Nietzsche was five. The young Nietzsche was raised by his mother, sister, grandmother, and two aunts. His early studies of the history of the Greeks, particularly the nobles in the Golden Age of Athens, greatly impressed him. These nobles lived only for the present moment and could accept victory or defeat with equal composure without traces of guilt. This may explain why throughout his life he kept two distinctions in mind—noble and slave, man and woman. Robert J. Ackermann, a professor of philosophy at the University of Massachusetts points out in *Nietzsche*, (1990), "In his [Nietzsche's] own view, decadence begins with the blurring of these distinctions."

Nietzsche elaborated on Darwin's thesis of biological evolution and believed that the moral conduct of the Judeo-Christian ethic was a "slave morality." He claimed that this morality permitted the weak to limit the self-realization of the strong. Nietzsche viewed men who were self-confident, assertive, outstanding achievers and were usually, therefore, highly visible as *Übermenschen*—supermen. They had reached a superior level of human development. Nothing could be further from Jesus'

statement that, "the meek shall inherit the earth." The meek were prac-
tically invisible.

Christian morality created what Nietzsche called a "herd instinct"
among these invisible people. Not much could be expected of them. In
contrast, his *Übermenschen* stood out high above the crowd in their
abilities in music, art, literature, philosophy, as well as warfare. Such
men would construct their own laws of morality and take responsibility
for them and therefore never experience guilt.

Kelly and Tallon (1967) called Nietzsche's philosophy, "biological pragmatism." A biological pragmatist who lived in an age in which we could
manipulate genes would have no reservations about using genetic engineering to redesign people who were permanently trapped by "slave
mentalities." A biological pragmatist might also recommend gene therapy
for clergy, who Nietzsche claimed used the Judeo-Christian values to
give people guilt in order to hold them hostage to the church. Nietzsche's
rejection of metaphysics and the social ideas of the 19th century caused
him to view the whole moral world order of his time as an invention
"against the emancipation of the man from the priest." If he could have
looked ahead to the Age of the Genome, he would have thought that
nothing could be as great a disaster as religious leaders having decision-
making roles in something as important as the application of gene technology.

Nietzsche viewed himself as a scientist and a philosopher, but actually he resembled a crusader most of all. When he wrote *The Antichrist* (1888), he called Christianity "the greatest misfortune of mankind
thus far." He also wrote: "In the whole New Testament, only Pontius
Pilate the Roman governor commands respect." We have to ask ourselves, what becomes of morality when the Judeo-Christian tradition,
the repository of morality in the Western world, is viewed as a misfortune?

To some extent, one could question the intent of Nietzsche's works
because his thoughts touched on many diverse subjects. His ideas were
close to thinkers of the Enlightenment, as well as to the romanticists,
logical positivists, existentialists, atheists, and sociobiologists. Some re-
viewers of philosophical thought consider him one of the most provoca-
tive and influential thinkers of the 19th century. A number of these view
him as a moral philosopher, since he consistently attacked hypocrisy,
indolence, and intellectual cowardice. After his death, his writings were

distorted and used to support Fascist and antisemitic world movements to which Nietzsche would have objected. His ideas were misused by the Nazis to promote the ideas of racial superiority of the Aryan race and justify the persecution of non-Aryan minorities. Kaufmann points out that Nietzsche made favorable references to the Old Testament and to the Jews of his time. In his writings, Kaufmann also notes that, "Any attempt to pigeonhole [Nietzsche] is purblind."

Kaufmann made a surprising statement when he wrote, "many Christians feel they understand [Nietzsche] best." How can one explain this in view of Nietzsche's bitter denunciation of Christian theology? Perhaps this was because Jesus believed that the rigid religious laws of Israel were out of step with human needs and Nietzsche thought the same of the religious theology of 19th century Europe. Jesus' declaration that "The Sabbath was made for man, and not man for the Sabbath" (Mark 2:27) appears to be at odds with the Roman Catholic Church, which has over two thousand canons to regulate every aspect of Christian life. The Church seemed to resemble a spiritual offspring of the Pharisees rather than the spirit of Jesus, who spoke out against the strict and literal adherence to religious laws. Had Jesus and Nietzsche lived in Spain in 1478, there is no doubt that both would have been called heretics and put to death by the Inquisition.

Even their sentiments regarding children were similar. Both thought of children as future actualizers of the potentialities of the universe. Mitchell, in *The Gospel According to Jesus* (1991), described the large crowd who waited for Jesus' arrival in Judea after he left Galilee, including children who were brought to be blessed by him. The parents were rebuked by the disciples for having brought their children, but Jesus told them, "Let the children come to me and don't try to stop them; for the Kingdom of God belongs to such as these." Nietzsche wrote in *Thus Spoke Zarathustra*, "A child is innocence . . . a new beginning . . . a first movement, a sacred 'Yes.'"

Nietzsche's Supermen excelled in overcoming and can-do. Jesus would have had nothing against the idea of supermen as far as their talents were concerned. However, he would have translated the word correctly as superpersons, as in the original German, since throughout his ministry he respected women and their rights more than was usual 2000 years ago in Israel. Jesus believed that all people, including Nietzsche's "sheep" were potentially *Übermenschen* and could rise to the highest level when using the power of their own inner visions. He would have disapproved of the supermen Nietzsche described because of their pride, feelings of superiority, and view of themselves as beyond sin and guilt.

Ironically, Jesus and Nietzsche agreed that the motivation of many militant supporters of their own religion whether Christianity, Buddhism, Hinduism, Judaism, Islam or others, stems from a desire to be right, instead of from genuine religious feelings. For such people, religious militancy becomes a means for self-assertion and power. It does not matter which religion they espouse, they use it to satisfy their need to declare everyone wrong who disagrees with them. In doing so, they do not seem perturbed that their self-rightiousness may violate the basics tenets of their religion, which typically include humility and respect for others.

This was well expressed in 1820 by Charles Caleb Colton, a noted clergyman and writer on religion who observed that "Men will wrangle for religion; fight for it; die for it, anything but live for it." The people Colton had in mind are numerous and vociferous today. They ignore the fact that the majority of members of a particular religion were born into a family with that religion. Their insistence that their religion is right stems from an accident of birth. Most did not select it after conducting an unbiased survey of other religions or philosophies. Self-righteousness and hypocrisy is what Jesus encountered among the religious leaders of his time and he, as well as Nietzsche, spoke out against them. Today, as in Jesus' time, it is dangerous to do so because militant religious people have been known to break the Ten Commandments as well as Buddha's injunctions against violence in the name of their religions.

It is the difference between the views of Nietzsche and Jesus that will be critical to people living in the Age of the Genome. Jesus had little hope for the proud and self-assertive who considered themselves above good and evil. Jesus would have believed that the people described by Nietzsche as a herd of sheep could more easily gain entrance to the Kingdom of God than the supermen so much admired by Nietzsche.

Nietzsche's theme—the drive to power—could be seen as representing the flow of the dynamics of the universe. When driven by pride, it could lead to ever greater accomplishments, acquisitions, and gains. William J. O'Malley points out in *America* (1994) that "Jesus did not consider wealth a sin . . . " His concern was not with wealth itself but rather the exploitation of others often associated with the acquisition and maintaining of wealth. Jesus taught that the flow of the dynamics of the universe should be directed to compassion and service without ulterior motives. As mentioned in Chapter 6, the 18th century philosopher Immanuel Kant's perception of morality echoed this. In spite of 2000 years of exposure to the teachings of Jesus and the Eastern sages that condemned pride and acquisitiveness, some people continue to act as if they were disciples of Nietzsche. This occurs to me every time I hear the stirring and triumphant words and music of "Onward Christian Soldiers, marching as to war, with the cross of Jesus going on before"— so like Nietzsche and unlike Jesus.

With regard to the meaning of life, Jesus and Nietzsche differed tremendously. For Nietzsche, God was dead and a Kingdom of God had no meaning. Jesus advocated transcending life's problems peacefully. Nietzsche chose to overcome them by meeting them head-on with courage. Transcending offers humankind an alternative that requires even greater courage. Resisting the use of power to manipulate our genome may take the greatest courage of all.

The message that Jesus brought to the people could mean that a can-do world alone could not raise humankind to a level that offered life's deepest meaning. There is nothing wrong with achievement in itself and transcending was not intended to replace it. The life of historical Jesus was, in itself, a magnificent achievement. We could see it as the actualization of a remarkable potentiality of the universe that could lead life into a new direction and does so when we adopt a morality to guide us. Nietzsche recognized Jesus' achievement even though he rejected Christianity. Surprisingly, Nietzsche included the name of Jesus on his list of supermen.

Beyond can-do, people require transcending to bring out the gentler qualities they also have within themselves. Jesus did not hesitate to struggle with those who violated a moral principle, as did those who exploited the sanctity of the temple. Forgiving must not be confused with overlooking. When Jesus used the fig tree bearing fruit to illustrate value of human life he struck a blow against racial and other types of prejudice. As quoted in the Gospel of Matthew (King James Version) Jesus taught "by their fruits ye shall know them." The period that follows the words represents a primary requirement for a genuinely moral world. The period aborts any addition to the phrase that begins with "except."

O'Malley believes that the clearest insight into the treatment of sinners by Jesus is found in the story of the Prodigal Son (Luke 15:11–32). A son left home with money that his father gave him and soon frittered it away. Later, he realized his mistake and returned home with a confession on his lips. His father welcomed his return with a big party. His brother, who remained home, sulked at this elaborate welcome. The parable teaches that we need not be punitive. The celebration was not only for the son's return home, but also because he recognized his error. His father took the Big Picture view of the event and did not permit a fragment of reality—his son's lack of good judgment—to represent the whole of his son's reality. In genetic engineering the Big Picture view must be kept in mind. It is that of the universe and that of humankind and not confined to the view of the gene.

Another of Jesus' parables brings this out clearly. He likened the Kingdom of God to "a mustard seed, which is smaller than any other seed: but when it grows up it becomes the largest of the shrubs and puts forth large branches, so that the birds of the sky are able to make their nests in its shade" (Mark 4:31). This parable suggests that transcending involves a change from the small picture view of the seed of self-interest to the Big Picture view of compassion and responsibility for all living creatures. The Big Picture view grows outward from within us and, since we are only the small picture, we resemble the smallest of seeds. A seed within us contains the potentiality that gives rise to a Kingdom of God. It transcends the insignificance and limitations of our human nature to branch out and become "the largest of the shrubs." Transcending brings into play a relationship between the self and the world. Under those circumstances, our genome may remain relatively safe from exploitation in the 21st century. In his enthusiasm for his Supermen's ability to express their assertive feelings, Nietzsche failed to appreciate the significance of transcending in the way Jesus advocated. Supermen or supersalesmen may well be the ones who will make the major decisions on how gene technology will be applied in the Age of the Genome. Would they resist the temptation to use genetic engineering or genetic cloning to perpetuate themselves? In that case, meekness is certain to disappear as a human characteristic.

Nietzsche offered four directives to his readers: (1) don't remain like animals; (2) don't listen to the moralists; (3) decide right and wrong for yourself and (4) seek eternity by creative efforts. He believed that the common people he called "a herd of sheep" would not be able to follow these instructions. It shocked his readers because he presented his views with forceful rhetoric at a time when the democratizing winds of the Enlightenment had swept over western Europe. Nietzsche's distinction between two categories of humans—the supermen and the herd of sheep—caused many people concern. I have already mentioned that the Nazi and Fascist dictators used Nietzsche's classification of people as inferior or superior to justify their actions. This could not have occurred had they accepted Jesus' view of humankind. Communism's avowed objective, "From each according to his ability; to each according to his need," seemed compatible with teachings of Jesus. However, when this idealistic goal was enforced by a dictatorship, it lacked a requirement of morality—individual freedom of choice. The failure of Communism illustrates that the best intentions in the world may, if corrupted, obtain the worst results.

The Crucifixion of Jesus has been interpreted in various ways. Cullmann, described as "one of the world's distinguished New Testament scholars," writes in *Jesus And The Revolutionaries* that it is "acknowledged by the majority of the scholars . . . that the legal condemnation was pronounced by the Romans rather than the Jews." He points out that the Romans were not interested in the religious squabbles and controversies of the Jews. They acted only on what they saw as a political threat to their rule. On the cross, as was customary, the Roman's posted their verdict, "The King of the Jews," which they judged as a political threat. Mitchell in *The Gospel According To Jesus* sums it up with: "It is clear that the Romans executed Jesus as a dangerous revolutionary."

Jesus died too young, Nietzsche lamented. He wrote in *Thus Spoke Zarathustra* that Jesus "himself would have recanted his teachings had he reached my age." In *The Antichrist*, Nietzsche viewed Jesus' death as a "guilt sacrifice" and added that it was the "most barbaric . . . sacrifice of the guiltless for the sins of the guilty!" He called the glorification of the whole scenario of Jesus' death and resurrection a "ghastly paganism!"

The Marginal Jew is a term that Father Meir, a Catholic priest, used as a title for his book on Jesus (1991). Meir described him as disgraced, flogged, and ridiculed while on the cross. Transcendence from there to the role of the Son of God provides a vision of truly heroic proportions. Nietzsche failed to see that the story of Jesus' death and Resurrection, whether it was based on actuality or a legend, represents a concept of the near limitless power of transcendence.

The phenomenon of Jesus personifies the "good news" that almost anything is possible for humankind—not merely for supermen. Contrary to what Nietzsche viewed in Jesus' resurrection, it fitted the parable of the mustard seed "which is smaller than any other seed: but when it grows it becomes the largest of the shrubs and puts forth large branches." Nothing written by Nietzsche matched this Big Picture of humankind.

Yet in a way that Nietzsche may not have intended, he was correct in thinking that something relating to the death of Jesus resembled "ghastly barbarism." It fits the description of Christendom using the Crucifixion to justify the hate of the Jewish people, to whom Jesus himself belonged. There are ardent Christians who claim to revere Jesus among those who have made a mockery of his mission by persecuting the people Jesus would have protected. The distortion of his message leading to anti-Semitism throughout the centuries symbolizes his ongoing Crucifixion. It was not until 1965 that the Roman Catholic Church formally exonerated the subsequent generations of Jews from their alleged responsibility for the death of Jesus. However, by 1965 the exoneration was too late. Anti-Semitism had already spread throughout the world and taken root, which enabled Christians as well as non-Christians to use it to enhance their perception of themselves.

We are still too close to it to fully recognize the irony of Christian-fomented anti-Semitism. Two thousand years from now, if we survive that long, students of ancient history who look back at us from the Big Picture view that time sometimes provides will shake their heads in disbelieve that Christians who accepted a Jew as the son of their God could be antisemitic, or tolerate it, knowing that his mother Mary, his brothers and sisters, his disciples, and the first missionary of the Christian religion, Paul, were Jews. From the Christian point of view, these thoughts are reflected in the following verse,

> Since it was God's wish
> To have a son who was Jewish
> It isn't at all odd
> That a dislike of Jews reflects on God.

Since God, Himself, created races
With different skins and different faces
Bigotry in any nation
Is against God, an accusation.

In Jesus' day, there was the promise of the Kingdom of God for those who transcended hate and prejudice and replaced them with love, acceptance, and compassion. We can only hope that in the 21st century the fear of the misuse of genetic engineering will cause people to adopt the kind of morality that the philosopher Immanuel Kant described. Perhaps we need this fear as well as the promise of a Kingdom of God on earth to put the teachings of Jesus into practice.

Records show that at the age of 45 Nietzsche became mentally ill. Let us imagine he suffered a disturbing hallucination in which he saw himself in the country of his birth, standing on a rise that overlooked a collection of grim-looking prison buildings. In a cluster of trees outside the compound stood the crematorium. Nietzsche noticed a uniformed guard below in the prison yard look up and call out to him, "Nietzsche, we are doing this in your name." Horror gripped him and he turned away. Not far from where he stood, he saw a young man weeping, his eyes on the crematorium below. Nietzsche wondered who he might be. When he heard his despairing cry, "My God, my God, why have you forsaken me?" Nietzsche knew who he was.

In the next century, Jesus may not, in anyone's fantasy, appear at Dachau to weep. Instead, he might roam the earth to mourn for an extinct humankind replaced by a new super-species—*Homo geneticus*—designed by Nietzsche. On the other hand, let's not totally rule out the possibility that Jesus might find that human beings had successfully adopted a new mind-set without resorting to genetically engineering themselves. Let's hope that this mind-set enabled them to transcend greed and turn the other cheek to hate. In that case, we could picture Jesus saying happily to himself, "At last the Kingdom of God has arrived on earth. It's a bit late, but it's doing well."

In the next chapter we shall explore the origin of free choice and it's role in human lives. The question we must ask and try to answer is, does free choice really exist or is the idea an illusion? The impact of the answer to this question on the direction genetic engineering will take is great. If free choice does not exist, we are already *Homo geneticus*.

CHAPTER ELEVEN

THE CONUNDRUM OF FREEDOM
OF CHOICE

Freedom of choice, sometimes called "free will," is usually considered a topic for discussion in philosophy and theology. When mental health specialists testify in criminal cases to help determine whether a defendant has a valid plea of insanity, it becomes an issue in psychology.

If it can be shown that the defendant in such cases could not have distinguished right from wrong behavior he would, by law, be declared innocent. He would not then be sentenced to serve time in jail but would be sent to a mental hospital for treatment. This seldom occurs. Even people suffering from mental health disorders usually retain sufficient freedom of choice to avoid criminal behavior.

The question of freedom of choice and free will has implications for the behavioral sciences and especially for psychology. We cannot isolate behavior from the larger context of human existence. The professions require us to fragment reality to enable us to study the particulars of the disciplines. Later, however, Humpty Dumpty must be reassembled in order to get the Big Picture. Otherwise, members of a profession limit their freedom of choice by viewing the world only through the lenses of their own disciplines.

In this chapter, I use freedom of choice and free will interchangeably, since whatever applies to one applies to the other as well. Freedom of choice has two parts: a) the condition which permits us to make choices that are independent of compulsions; b) the knowledge that alternate courses of action are open to us. If these conditions are met, we are not forced to reach predetermined conclusions. The *Dictionary Of Philosophy* uses three words to define freedom—"self-determination, self-direction, and self-control."

We cannot claim that freedom of choice exists exclusively for humans. Animals also have some options as to whether to feed here or there, whether to fight or flee, or to risk offending the dominant male or female of their group. To some extent, they also have some freedom in mate selection. In this chapter, I refer to the freedom of choice and the mental experiences humans can have in the area of abstractions and intangibles. Among these is the ability to accept or reject a freely accepted morality, to adopt a humanitarian sense of justice, to consider the existence of an afterlife, as well as other questions of meaning and significance.

No exact date can be set for the time in prehistory when humans gained the ability to make abstract free choices. Our knowledge of the chronology of prehistoric developmental events is limited. To use the size of fossil skulls as a sign of the ability to think is misleading. Even today, some mentally handicapped individuals have larger brains than people of normal intelligence. Complexity, neurochemical factors, and density of neurotransmitters are some of the characteristics of the cortex that distinguish the normal human brain.

Some paleontologists believe that signs of hominid ability to think abstractly may have existed as far back as fifty to sixty thousand years ago. In a cave called, "Shanidar" in Iraq, Neanderthal skeletons of a man, two women, and an infant were found buried together with the pollen of flowers which could not have grown inside the cave. Some anthropologists infer from this that some kind of a ceremonial burial may have taken place among the Neanderthals. Flowers placed at the grave suggest that the Neanderthal people had rudimentary concepts of an afterlife. We cannot rule out that they may have had at least a flicker of spirituality.

Questions about life and death must have occurred early in the creative evolution that later led to the concept of free will. Any people who had gained even the slightest notion of freedom of choice would realize that it was strictly limited by life's circumstances. External events, the powerful influence of other on us, our ethics, as well as our own emotional or mental limitations diminish free choice—sometimes to the vanishing point. In the forefront of the limiting factors is lack of knowledge. It is self-evident that the less knowledge we possess, the less free choice we have.

This is analogous to the "surprise gift" packages that mail order houses occasionally offer their customers as incentives to purchase a certain quantity of merchandise. Surprises are welcome, but if we don't know what a surprise package contains, we lose all freedom of choice in selecting something that we want or need. Long ago in ancient Greece, Plato said that free will is limited by the extent of a person's knowledge.

In spite of the bitter battles people have fought throughout history to obtain freedom of choice, once it is gained many people do not necessarily welcome it. Free choice is a double-edged sword. It requires us to take responsibility for our decisions and that can be disturbing. We recall from the previous chapter that Nietzsche viewed many humans as sheep without free will. He referred to people's dependency on their priests and their religion for decision-making. It is true that it is easier for sheep to delegate responsibility to their shepherd than for them to use freedom of choice themselves. Some people speculate that sheep have a more relaxed life than we do.

Even under circumstance favorable to the expression of free will, it is restricted among people struggling to secure the bare necessities of life. Likewise, it is curtailed by feelings of anger, by disease, and emotional turmoil. As handicaps of all kinds increase, freedom of choice often decreases proportionally. But there are exceptions. Some people rise above their circumstance and disadvantages. Instead of losing free choice, they gain an additional amount of it by not permitting their limitations to hold them back.

Paradoxically, we often make little use of things we fought hard to obtain. An example is the large number of people who do not vote in a country in which freedom of choice was gained through great sacrifice. Free will is endangered by the fact that it may distract some people from a life lived in comfortable niches they have carved out for themselves. One of the "advantages" *Homo geneticus* would have over us, if we let this genetically automated species replace us, is that they would have no need to make decisions. If we think about it, isn't that also a major difference between life and death?

In time, geneticists may be able to program behavior in people's brains like cardiologists now program the performance of the pacemakers they implant. Even today, freedom of choice is only viewed as a virtue in a democracy. The first casualties in a totalitarian regime are freedom of thought and choice of religion. Dictators know that they must suppress both of these, since in Western religions God is the supreme power and that puts the dictator in second place. Absolute rulers either represented themselves as an embodiment of gods or ruled by "divine right."

Barring this, dictators must abolish free will since it is the cradle of new mind-sets that often trigger discontent. Unfortunately, history has shown that religions can be manipulated to serve as tools of dictatorships. When they do so, they are self-destructive. Without freedom of choice, sin would have no meaning in Christianity, God's anger at human disobedience as related in the Old Testament would be idiocy, the Buddhist goal of escape from individual existence and the attainment of Nirvana would be foolishness, the Hindu idea of progressively more blameless lives leading to liberation would be science fiction, and prayer of any kind would be equivalent to baby talk.

Without freedom of choice, every act of punishment would be unjust. Jails would represent the epitome of injustice. Even criminals caught red-handed would have to be declared innocent by reason of an absence of freedom of choice. In a deterministic world, everything that happens would be inevitable, and the only sensible view of life would be fatalism generously mixed with hedonism, as we discussed this in Chapter Six. To some extent, this represents the world in which we live today. Many who perceive an absence of free choice in their lives succumb to alcoholism and drug addictions or find other ways of escaping from an unsatisfactory life.

To give life meaning, free choice must exist. The sticky point is that some persons' free will calls for other persons' surrender of theirs. As I pointed out in Chapter Nine, this did not disturb the English philosopher Herbert Spencer, whose theory of *laissez faire* is well named—let things happen without interfering. At first sight, noninterference appears to maximize the opportunity for freedom of choice. It claimed to lead to morality, since it was seen as the ultimate test of freedom. Unfortunately, it suffers from one overriding flaw—other values beyond surviving would drop by the wayside. Morality would be defined according to laws of the survival of the fittest, leaving the definition of the "fittest" unchallenged. Thus, *laissez faire* limits the freedom of choice, even for those who come out on top of the heap. The principle of *laissez faire* offers them only a monolithic criterion of success. Free will requires individuals to be able to select their own definition of success, which may vary greatly from that produced by *laissez faire*. Nietzsche's perception of success did not resemble that which Jesus taught when he declared: "The meek shall inherit the earth." To Nietzsche, this would have appeared as the greatest failure imaginable.

When psychologists place self-actualization among the human needs, they mean that we must dig into ourselves and bring out something that lies within us. Self-actualization requires us to make choices that are independent and voluntary. Choices always result in actions, while external events and instincts only produce reactions. Reaction is the raw stuff of which Darwinian evolution is made. This does not hold for creative evolution, which might be viewed as an outgrowth of biological evolution. Darwinian evolution no longer applies to humans. We acquired the mental equivalent of mutations by means of the thought processes in our brains. These thought processes have led to our unique human free will, in which we differ from other forms of life.

Stephen Jay Gould, a prolific writer and professor of the History of Science at Harvard University wrote in *Ever Since Darwin* (1977): "We are both similar and different from other animals." In his opinion, "It is impossible to overlook the extent to which civilization is built upon a renunciation of instinct." Instincts are reactions to inner promptings devoid of cognition, planning, and freedom of choice. A baby bird taken from its habitat and all contact with its own species at birth will, nevertheless, build an identical type of nest as other members of its species. Actions, on the other hand, allow for innovation and creativity. New mind-sets created by reactions may serve survival, but only actions can lead to self-actualization. Animals are capable of learning and improvising and even of originality. Not all animals are totally instinct-bound. However, as Gould points out, humans differ from all other forms of life by "the extent . . . of our renunciation of instincts." Humans are able to replace these with independent judgment and freedom of choice.

We have reached the point in our discussion of freedom of choice where we must confront a challenge to its very existence. Robert Wright, in his book *The Moral Animal* (1994), calls free will "a fairly useful fiction." We cannot seriously consider the attributes of something that may, in fact, not actually exist. No one has ever been able to prove that freedom of choice exists. The idea of free will rests on very shaky ground. Attempts to prove that it exists usually consist mostly of wishful thinking. It can be argued that actions we consider as free choices could, in fact, be forced upon us unknowingly by previous experiences and inherited tendencies. The theory of evolution gives us the impression that free will was not included in nature's grand scheme for life. Organisms survive because randomly produced mutations help them adapt to environmental demands. In spite of its obvious success, evolution appears to be an impersonal hit-or-miss process. The survival of the fittest is often based largely on luck in the dice game of errors in the replication of genes. This would make to the perception of free choice nothing more than an illusion. The entire theory of evolution enabled by random mutations speaks out against free will at every level of life and especially at the level of bacterial life. The physics teacher I had in high school taught us (in somewhat poetic language) that "nature abhors a vacuum." We might also deduce from random selection that nature abhors freedom of choice.

However, in a 1994 issue of *Science,* in the section devoted to research news, Elizabeth Culotta wrote an article titled, "A Boost for 'Adaptive' Mutation." The article begins: "British geneticist John Cairns touched off a firestorm in genetic and evolutionary circles . . . by proposing that bacteria could, in his words, "choose" the mutation they acquire. Assigning such freedom of choice to bacteria violates the basic tenet of evolutionary theory, which holds that mutations arise at random."

Cairns and his colleagues had found evidence for their view when they put bacteria unable to digest lactose into a petri dish with only lactose for nourishment. The researchers found that bacteria preferentially acquired the crucial mutations they needed to become lactose-eaters. The article continued: "Then came the firestorm. Other researchers also saw evidence of these apparently nonrandom mutations in their accounts of various bacterial and yeast colonies."

The article reported that a "legion of neo-Darwinists and population geneticists" considered Cairn's idea a heresy of the first order. They pounced on his work, charging that its results were due to "experimental artifacts." In other words, there was no free choice! Yet other biological scientists admitted that something was going on. In no way could they dignify it by calling it free choice, but they described it as "Selection Promoted Additional Mutations" and suggested the best thing to do under these embarrassing circumstances was to call the phenomenon, "SPAM."

Random chance was somewhat tarnished by SPAM, but it won out. The German 18th century philosopher Leibnitz held that things that one would logically expect to happen do not necessarily happen, while things that seem illogical do happen. According to Barrow and Tipler, who wrote *The Anthropic Cosmological Principle* (1986), the chance that the combination of gases and elements exactly in the precise amounts necessary for life would exist anywhere in the universe was extremely unlikely. One of the authors suggested later that it was more plausible to account for the existence of life by supernatural design than by random chance alone.

Examples of the unlikeliness of random chance occur more often than people think. For example, President Lincoln was elected in 1860, President Kennedy in 1960. Lincoln's secretary was named Kennedy; Kennedy's secretary was named Lincoln. Both presidents were shot on a Friday with their wives present. Lincoln's assassin shot him in a theater and hid in a warehouse; Kennedy's assassin shot him from a warehouse and hid in a theater. Both assassins were themselves killed before their trials. When coincidences of this kind occur, we are apt to suspect that something beyond random chance occurred. Perhaps that is why the English 18th century philosopher F. H. Bradley decided that chance and free will are indistinguishable. If true, no one can ever prove that free will exists.

Let us look elsewhere for an answer to the question of freedom of choice. A Tibetan Buddhist of the 20th century, Chogyam Trungpa, wrote in *The Myth of Freedom* (1988): "We are never trapped in life, because there are constant opportunities for creativity and challenges for improvising." There would be no place for improvising or challenges if life consisted of certainty. All arguments in favor of freedom of choice hinge on the premise that uncertainty exists in the universe. Without uncertainty freedom of choice could not exist. Free will is difficult to identify because it merely converts uncertainty into less uncertainty. In every case, some uncertainty remains. An example might be people who freely commit themselves to a lifetime of partnership, as in marriage. Uncertainty as to whether or not they will live their lives by themselves diminishes. However, one of the partners may die or otherwise leave the partnership. Freedom of choice did change the odds, but failed to create certainty. One may be discouraged at the limited power of free will, but nevertheless, reduction of uncertainty is of immense significance and invalidates the exclusive rule of random chance.

Those who believe in supernatural powers claim that God gave humankind free will and therefore without further fanfare or tortured reasoning they believe that free will exists. Science is unable to offer anything as logical and forthright as this deductive reasoning. However, the view is restricted to those who accept the original premise. Even this straightforward solution creates a quandary in theological circles. It is called, "The Paradoxical Problem Of Free Will." The dilemma is one of logic. If God is omniscient, He has prior knowledge of every choice that each human being will make. In this case, humans cannot freely choose or act in others ways than those of which God has advanced knowledge. If people could act contrary to God's knowledge, God would not be omniscient. The paradox continues with the proposition, if God has foreknowledge of everything and is omnipotent, He must have organized events in advance of their actual occurrence. In that case, humans cannot be said to have free will. They must follow God's plan.

In the 13th century, Saint Thomas Aquinas, the Italian philosopher and theologian who wrote the monumental *Summa Theologica* easily brushed the problem aside. He declared that God's omnipotence does not include predetermination of human will because God did not wish it to do so. Since God is omnipotent He can exclude whatever He wishes from his own omnipotence. God wished his children to learn something on their own. In essence, God is seen as having the will, at times, to look the other way.

Those who cannot accept such supernatural explanations confront the thorny issue of cause and effect. How is it possible to reach a logical conclusion in view of the fact that we accept the law of cause and effect—stimulus and response? This law can only lead to the conclusion that we live in a mechanistic universe, since anything we call free will must have been caused by a stimulus and cannot, therefore, be viewed as free will. In science we face the theological paradox translated into non-theological terms. Some would refer to the work of the Norwegian physicist, Niels Bohr (1885–1962).

Bohr's philosophy and quantum theory, mentioned in Chapter Eight, suggest that subjectivity enters into the observed behavior of subatomic particles. This and other new discoveries outdated the idea of a mechanical universe in which all events would be predictable. Heisenberg, whose Uncertainty Principle we discussed in Chapter Five, proved that uncertainty exists in how humans are able to view subatomic particles. A vexing questions arises. Is Heisenberg's Uncertainty Principle the result of human inability to be at two different locations at the same time? The arguments for freedom of choice provided by physics are no more tenable than those from theology. Both rely on faith.

Therefore, faith has to exist in human life whether we believe in theology or science. Free will relies on faith and our very humanity relies on free will. In Chapter Six, I stated that creative evolution allows for a bit of craziness within us. Faith without supporting evidence is the kind of craziness that saves our sanity. This leads us straight into the hornet's nest of existentialism which claims that even thinking puts us into a position of unbearable responsibility. Some existentialists tell us that the awareness of our responsibility resulting from our freedom of choice overwhelms us. However, critics of existentialism deny that we have such a vast amount of free choice. Most thinkers realize that free will without responsibility creates tyranny. At a time when existentialism captured the minds of many people, Mary Warnock, in her book *Existentialist Ethics* (1978), made the limitations of freedom clear when she wrote: "If freedom for oneself is the highest value, the free choice to wear red socks is as valuable as the free choice to murder one's father or sacrifice oneself for one's friend. Such a belief is ridiculous."

It seems that we may have conjured up a dangerous monster through our claims to have unrestricted freedom of choice, as "murder one's father" implies. Free choice exists only when there are alternatives. This must apply to goodness—hence Satan or evil must exist. In providing alternatives to goodness, we make freedom of choice possible. It offers human life a purpose which is to overcome evil, banish it, destroy it, replace it, or at least, diminish it. Seen from this view, the existence of evil and cruelty make sense. Without alternatives, morality would have no meaning. It is ironic that without the opportunity to choose between good and evil no one would be truly free. By personifying evil, Satan gives pursuing goodness its important meaning. Goodness, as Eve learned to her dismay, consists of avoiding temptations. In this there is an important lesson for those who may be tempted to approve of unrestricted genetic engineering.

Richard P. Feynman, who won the Nobel Prize in physics, said that he felt vastly better about himself when it dawned on him that he was not personally responsible for all the things that happened in the world. The art of living must, in part, consist of knowing our limitations. I saw it all summed up in a fortune cookie I opened last night after a dinner in a Chinese restaurant. As I broke it open, the fortune cookie expelled a small piece of paper which stated (in capitol letters), "WE CANNOT

DIRECT THE WIND BUT WE CAN ADJUST THE SAILS."

To a large extent, free will is something that each person must give to or give up for him or herself. In his book *Freedom And The Moral Life* (1969), John K. Roth reviewed the ethical thoughts of William James. In the preface he wrote: "James believes that our lives are permeated by a freedom that gives us a chance to shape the world we inhabit." Further along in the Preface he strikes a key note. He asks: "If I am free in a variety of ways, how shall I act, and which values should I take to be the most important?" Those who attain freedom will always encounter this question.

In the chapter that follows, we shall search for a new mind-set appropriate for the time when gene technology will play an increasingly larger role in human life. We are living at a time when genetic engineering does not yet press itself upon us with sufficient force to render us helpless. We still have an opportunity to build our future. This requires freedom of choice. Whether we rely on scientific fact or faith, we must believe that freedom of choice exists. Without it, we would not have any role in selecting a new mind-set. This is the topic of the next chapter.

CHAPTER TWELVE

A NEW MIND-SET

A new mind-set follows every major change in the way we cope and live. More than a million and a half years ago, our early ancestors gained a new mind-set after they fashioned crude tools and weapons of stone and wood. Chipping stones created a new mind-set early in human prehistory and so did the use of fire and, later, the invention of the wheel. A significant change in mind-set occurred about 12 thousand years ago, after some hunters and gatherers abandoned their nomadic lifestyles and turned to agriculture and domestication of animals to assure them of a reliable food supply. In good years they could accumulate surplus food to tide them over in leaner years.

The most significant new mind-set in all of human history will be created in the lifetime of many people who are alive today. No event in the past compares with the awesome power over nature humankind will acquire when the Human Genome Project is completed. A map locating the position of all the genes of any species will serve geneticists as a guide for redesigning the life of that species. In this chapter, we shall search for a mind-set that would be appropriate for this monumental change. Without our awareness that a mind-set is in the process of being created, we are unable to play a role in shaping it.

Let us begin with the research student. For that I refer to a recent *Center For Bioethics Newsletter* (Spring, 1996), published by the University of Pennsylvania. Glenn McGee, Ph.D., the newsletter's editor, and I agree that "scare tactics seem counterproductive to humanism in science . . . ." Nevertheless, history demonstrates that "scare tactics" are appropriate for scary situations that occur in the presence of apathy. McGee he notes that "The past few years have seen infamous cases of stealing, altering, cooking, erasing, and concealing data, involving millions of dollars and hundreds of patients." He suggested it might help to prevent this if "the research students' mentor" were more deeply involved. McGee sees it as the mentor's responsibility to inform students of rules and regulations that govern research with "radioactive materials, infectious agents, recombinant DNA, human or animal subjects. . . ." McGee correctly points out that ethical research could still breakdown unless students had a "sense of personal accountability for the character and meaning of science itself." Whereas scare tactics may not work, McGee feels that "personal accountability" could. Unfortunately, it is quite unlikely that any mentor could convince research students to become accountable who were raised in a culture where accountability is not practiced.

Accountability must be sufficiently emphasized as important in life from earliest childhood to adulthood. This cannot occur with our contemporary mind-set. Mind-sets are created by one's culture which is, in turn, shaped by how people of a culture regard the world's phenomena, including themselves. I refer to the theory of natural selection as a guide to human behavior. The time has come, and perhaps almost passed, for this theory to be replaced by one that projects a different image of ourselves. The theory of creative evolution offers us an alternative. Freedom of choice, discussed in the preceding chapter, enables some of us to discard a culture's mind-set, just as those world leaders who created major social changes with non-violent methods have done.

As we think of ourselves and our place in nature, it is well to remember that we can look at almost everything from two points of view. One of them focuses on an actual thing itself. The other view reveals its connection to the world around it. The first view could be called the small picture while the other one the Big Picture view. The small picture view has the advantage of highlighting details. The Big Picture gives meaning to what is looked at and relates it to the world in which we live. Our mind-set is formed by the view on which we focus. Small picture and Big Picture views lead to separate theories of origins. The small picture view considers beginnings in isolation from other events which leads them to become detached from the processes they bring into being. To understand the significance of origins we must go beyond them to speculate on their consequences, that is, see the Big Picture. The Big Picture projects an event, large or small, as part of the pattern of all of nature. The Biggest Picture view takes us, inevitably, to the nature of the universe itself. Our special concern as humans is, how and why the Big Bang (as the origin of the universe is colloquially called) has led to life and to ourselves. Our perception of where we came from creates our image of who we are.

Long before the Judeo-Christian tradition, creationism explained the origin of the universe in terms of the supernatural. Gods or spirits, separately or in unison, and later one single supreme God, were believed responsible for the existence of all things, including life. I shall not deal here with the supernatural, except to note that it represents the kind of a Big Picture view to which humankind is intuitively attracted. I must, as before, refer to the Darwinian view of life in order to demonstrate the extent of influence it has had in creating our contemporary mind-set. In contrast to creationism, Charles Darwin in 1859 focussed on the small picture view of life when he wrote the book, *The Origin of Species* and later *The Descent of Man*. In these books he startled the Western world with his theory of evolution. It left an indelible mark on the character of Western civilization which helps to account for some of our contemporary dilemmas.

From its beginning, Creationism relied on faith, whereas evolution explained the origin of species based on the theory of natural selection. Later this concept was described somewhat simplistically as the survival of the fittest in its focus on the ability of various species to adjust to varying environmental conditions. Darwin's work was based on keen observation and insight drawn from wide observation. The subsequent discovery of mutations and the notion of "selfish genes" helped give it credibility. Today's philosophical writings drawn from neurophysiology treat behavior as originating within the brain. Thus, with passing time perceptions of behavior became increasingly more narrow as they are ascribed to fewer and fewer origins. With neurophysiology seen as the basis of the mind and the soul, the small picture view of life became even smaller.

I have already presented the traditional view of evolution that spells out success in life as the reproduction of offsprings which, in turn, produce other offsprings, thus continuing the species. The key to evolutionary success is seen as adaptation and exploitation of ecological niches that lead to biological advantages. Put in the simplest terms and applied to human life this adds up to the idea that the world is out there for your benefit and you should grab as much as you can for yourself and for your species. Anything that does not touch upon this objective is peripheral and of not much concern. For humankind it would mean that behavior called "good," "decent," "unselfish," and "ethical" serves solely as a means for self-preservation and self-aggrandizement. However, because we also have a conscience, the selfish motive, according to the believers in the selfish gene theory, is often disguised. Self-centered views of life may be used to justify violence and exploitation. In saying this, I must point out that Charles Darwin would have been disturbed had he anticipated this interpretation of his theory. He was a well-intentioned man who had studied medicine and later entered Cambridge University, in preparation to become a clergyman. His primary intentions were to understand the origin of what he found occurring in nature. He lived at a time in Western civilization when there was a spurt of interest in searching for causes of natural phenomena.

After Darwin returned to England, he was influenced by the British economist Thomas Robert Malthus, who claimed that population growth would outstrip the available food supply. Malthus believed that in order to maintain balance there had to be natural limitations created by famine and disease or by social action such as war. Darwin applied Malthus's argument to animals and plants. In *The Origin Of Species* (1859), Darwin made this prediction: "We may look with some confidence to a secure future of great length. And as natural selection works, solely by and for the good of each being, all corporeal and mental endowments will tend to progress toward perfection."

Long before Darwin lived, ancient people viewed the "fittest" as victors in conquests and wars. They were the masters and the unfit, who lost wars, served them as slaves. Strength, and cunning led to the acquisition of power throughout recorded history. Successful use of force marked people who used it as favored by natural selection. Alexander the Great, the most successful conquer of ancient times, was on Nietzsche's list of supermen. Darwin, certainly, did not introduce the idea of the survival of the fittest. He did not associate natural selection with wars and physical conflicts as much as with peaceful adaptation.

The theory of evolution, intentionally or not, provided a yardstick for measuring success for those who survived at the expense of others. It brought the modern world a biological justification for competition and strife in the name of survival. This was not advocated by Darwin himself, but by some of those who interpreted him later. Natural selection is tragically misapplied when it is used to formulate the meaning of human life. When the time comes that we can effectively redesign ourselves, perceptions of ourselves drawn from biological evolution may prove disastrous. If human nature were forced into this kind of a mold, we would not be able to write poems, compose music, paint pictures, or be creative unless these activities served to further our chances of physically surviving or, as in ancient times, served to put us into the good graces of one of the gods. All behavior, including altruism, ethics, and morality would have to contain some underlying, perhaps camouflaged, self-serving purpose. Things done for themselves—merely for the sake of creating—would be seen as a total waste of time and effort.

Robert Wright (1994) supports the new Darwinian paradigm of the selfish human gene. He views open or disguised status-seeking as the basis for all human interactions. If we asked, "Why do we tend to disguise our status-seeking and selfishness?" some would reply, "In order to disarm others." It may occur to some of us that we reject status-seeking and self-centeredness because it fails to conform to our belief system. Who knows, before the human genome is fully mapped, molecular biologists (to their surprise) may discover a shy gene for unselfishness among our hundred thousand. Wright, who suggests that the Golden Rule itself is a product of natural selection would be inclined to doubt it.

Nevertheless, a species that has the capacity for use of imagination and abstraction may do some unexpected things as time goes on. For example, we might branch out further into creative evolution that would bring human attributes other than merely selfish ones into play. Although evidence of nature's diversification are all around us, modern evolutionists are stuck with natural selection as the one and only explanation of human behavior—past, present, and future. They confuse the push of the gene with the push the universe exerts to actualize its potentialities. The gene is only responding to the cosmic dynamics.

Because the theory of evolution lacks connectivity to the larger universe, it represents a small picture view of life. The problem is that it is used to reach Big Picture conclusions. For example, terms such as, "evolutionary pressure," found in some biology texts used to explain what keeps life changing and evolving, bypasses the question: Where did evolutionary pressure come from and why does it exist? To answer that question, we must turn our attention to origins. Evolutionary pressure will then be seen as a consequence of the pressure exerted by the universe's expansion from a tiny dot to our present cosmos. We realize that the universe is still expanding and in this we can see hope for the future of humankind. We are co-creators of the expansion process as, in our unique human way, we bring intangible qualities into being and add them to the universe's reality. Intangible qualities, such as morality, ethics, and compassion cannot be represented directly by images or described completely by words. The intuitive writers of the Old and New Testaments gave us this insight when they informed us that we must not worship idols or make images of God, or even name Him. Our self-actualization appears to require that we move from concretions to abstractions. Only then can we conceive things existing just for the purpose of existing. As we put our ideals into practice, we help the universe expand into a new dimension. The theory of evolution is incomplete because it fails to go beyond life on earth to find connectivity which, itself, is an abstraction. Creationism, no matter whatever else we may think of it, does not suffer from this defect.

Within science, it is also possible to obtain a Big Picture view of life—of humankind and of ourselves. I proposed this in a paper, "Speculating on the Roots of Human Behavior Beyond Biological Origins," presented at the Annual Meeting of the Pacific Division of the American Association for the Advancement of Science (1989). I proposed that, "The common ancestry of human behavior with all else that exists leads the search for human roots to the beginning of the universe." I viewed human nature as a derivative of the nature of the universe and suggested that human behavior may have obtained some of its characteristics from enduring patterns manifested throughout the universe in the physical world.

After the paper was reprinted, I received a note from John Archibald Wheeler, Department of Physics, Princeton University, who called the ideas expressed in the paper, "thoughtful and thought-provoking." John L. Casti, Professor of Operations Research and System Theory, at the Technical University, Vienna, read the reprint and commented, "Looking beyond heredity and environment for origins of mind and behavior, may well turn out to be the key that unlocks thorny issues." In the paper, I proposed that, "Maintaining balance in human affairs and in the lives of organisms is usually explained as resulting from natural selection for survival. But maintaining balance isn't what is really selective. Being in touch with the laws of nature that require balance throughout the universe is what is selective." Everything that exists is held together in one way or another by a connectivity that our minds tend to fragment. Fragmentation in our world outlook has carried over to how we perceive our fellow human beings. In the chapter on reality, we pointed out that fragments of reality conceived of as wholes struggle with each other to claim sole authenticity. The first step in developing a mind-set appropriate to the age of genetic engineering is to incorporate the small picture view of life into the Big Picture view of the universe—its laws, its processes, and its limitations.

In the physical world, actualized potentialities of the universe are manifested by forces of nature: the strong nuclear force, the weak nuclear force, the electromagnetic force, and gravitation. Physicists have searched for a fifth force, which they named a Grand Unification Theory (GUT). A number of candidates, including "superstring theories," have evolved. GUT represents the ongoing search for the Big Picture view in physics. In order to create a mind-set that can deter us from abusing genetic engineering, we need a similar effort in the human sciences. Diversification of life is not merely a product of biological evolution, as we have assumed. We can account for diversifications and varieties of organisms by ascribing them to the tendency of the universe to actualize its potentialities. Life is believed to have originated on earth 3.5 billion years ago after amino acids were formed into soupy chemical pools. These were, perhaps, energized by lightening strikes or solar radiation or, possibly, by matter from a striking asteroid. At its beginning, life may have consisted of little more than packets of the precursors of DNA. From there it has evolved to include the concepts of morality, justice, and compassion in a world that previously was governed solely by random chance.

We should not take too much credit. Both life and non-life participate as partners in actualizing the potentialities of the universe. Evolution, like other phenomena of nature, is a tool of the universe used to actualize its potentialities and maintain them once they have been actualized. The universe's tendency for innovation and conservation are mirrored in life—even in the politics of nations. Seen this way, randomness takes on a new meaning. Both random and designed events represent actualizations. Only a few of the large number of mutations that occur randomly lead to survival. From the small picture view of evolution, harmful mutations have been called nature's "mistakes." From the Big Picture view, they are tiny examples of the dynamics of the universe—

with its invisible hand—exerting pressure for actualizing its potentialities.

Jack Cohen and Ian Steward wrote, in *Discover* (April, 1994), "Different causes can produce the same effect." For example, different methods of flight in animal life have been "discovered at least four times in the history of evolution." Entirely different methods of sexual reproduction and determination of gender join the list. It seems that it is not as important how things are done as it is that they are done and that the maximum potentialities of the universe are actualized. These may be for better or for worse as far as life is concerned. The universe is neutral. Nature lacks compassion. Compassion—and a morality without ulterior motives—are potentialities of the universe humans brought into existence. They represent a major contribution to the make-up of the universe. We should be aware of this and teach our children to see them in this light.

In a mind-set appropriate for the 21st century, we shall recognize all the phenomena of nature as partners in the effort to actualize the universe's potentialities. This will bring us to a Grand Unification Theory that encompasses all of the sciences and sheds light on our role as citizens of the cosmos. The Big Picture arouses humility within us. This is a necessary quality as we enter an age in which we will have the ability to redesign ourselves and all of life. We must start searching for a new mind-set consistent with a Grand Unification Theory that includes all of the components of the universe we are able to recognize. Only then can we succeed in bringing humankind together into a similar reunification, as we must in The Age of the Genome.

It is imperative that a Grand Unification Theory in the human sciences replaces the small picture of life painted on Darwin's canvass. Our faith has been misplaced in the myth of "the most fit," championed by Nietzsche, whose philosophy we examined in Chapter Ten. We compared it with the Big Picture view we can discern in Jesus' parables. Nietzsche declared God dead, but he was wrong. God is alive, but has been captured and made the servant of a plethora of self-righteous ideologies and cults. He has been converted into a tool for the acquisition of personal and group power. The theory of evolution has been vigorously opposed by some religious groups who, themselves, have adopted religious Darwinism.

We must rely on religions as signposts that point the way to a mind-set appropriate for our times. But first the world religions must divest themselves of the religious Darwinism with which they are afflicted. Religion may have been the earliest manifestation of a rebellion against the tyranny of random chance. Religions differ, but all of them move beyond biological imperatives in their teachings. We can identify Christianity with humility, Buddhism with self-denial, Islam with submission to God, Judaism with justice, modern Hinduism with social concerns, Taoism with selflessness, and Jainism with reverence for living things. These religions, all of which predate our modern era, have each presented us with a new mind-set. But it is a worthless one if not put into practice in everyday living.

It took well over a million years for near-humans to gain our present *Homo sapiens* brain, and yet we are merely a little higher than animals if we measure our worth by biological standards of success in survival and reproduction. By that measure, we must in fact be described as a little lower than the insects. Only within a Grand Unification Theory alternately called the Kingdom of God can we become a little lower than the angels.

Archie Bahm, Professor of Comparative Religion at the University of New Mexico, wrote a chapter describing Hinduism in his book *The World's Living Religions* (1964) that states that some of the Hindu religious writings ask: "What does one actually receive when his action has been motivated by reward? Calculations about amounts of rewards tend to degenerate into distrust, cheating, quarreling, and, eventually, noncooperation."

In Chapter One, I quoted Karl Menninger, who described these conditions as typical of our present world. We ask again now, is this the kind of a world that is ready for gene technology? The answer is clearly no, unless we acquire a mind-set that can change no to yes. The key to achieving such a mind-set is not magical or recondite. It may be so apparent that it will disappoint those who look for something more spectacular. The mind-set that can change no to yes in continuing further research in genetic engineering consists of nothing more elaborate than finding a purpose in our lives beyond mere existence.

This takes us back to Chapter Nine, where we examined the roles of harmony and strife in nature. We reached the conclusion that both exist in the world. Both can be misapplied. The idea of harmony as dominant in nature is not convincing when we take into account the ongoing struggle of animals enacted every day in the food chain, in establishing territories, and in the competition for dominance among social species. We must not look to nature for guidelines we should adopt in the 21st century. They do not represent purpose for living beyond the one proposed by social Darwinism. Such guidelines are not appropriate for family life, for how we conduct industry, how we worship our God, what we teach our children, or what we watch on TV. A sense of purpose will place us "a little lower than the angels," instead of much lower than some bacteria that survive in temperatures near absolute zero and above the boiling point of water.

At the beginning of this chapter, I pointed out that every new experience—chipping rocks to make tools, the move from foraging to agriculture, the invention of the wheel, a human foot print on the moon—provided us with new mind-sets. All of our great inventions and innovations did this. However, if we use freedom of choice we can create a mind-set of our own choosing before, instead of after, our inventions have been applied and bring changes into our lives.

We might say that God gave us a purpose for living. In His wisdom, he left it for us to discover what it is. From its inception, the universe made purpose available to us as a potentiality. However, it was left to us to find it and actualize it. Because the discovery of purpose has been left to us, our appreciation of it will be greatly enhanced. The good news is that an earnest search for a purpose for living beyond survival suffices to create the new mind-set that is lacking today.

A poem in my collection *Poems For Living* sums up the main thought of this chapter.

THERE IS A PURPOSE IN THE WORLD.
Some people tell us that the world lacks purpose
"Life has evolved through blind chance," they say
Random mutations account for natural selection
Produced by adaptation to an ever changing earth.

In their myopia they view "the selfish gene,"
As the supreme monarch of all things alive
This single-minded monarch gives but one command,
"Survive, increase, spread me across the earth!"

Let us not believe that we know all there is
About purpose in the universe, but we can ask,
"Could it be that with the advent of human life
The universe found purpose that some fail to see?"

Where is the scientific evidence for purpose?
It lies unrecognized within that very question.
Purpose is found in asking who we are,
Where are we going, and what can we accomplish?

With humankind a purpose for the world was born
To banish random chance and replace it with justice
And to rebel against the rule of selfish genes,
To sing and dance, to play and to create beauty.

The world lacks neither purpose nor great goals
Because through human life it has acquired them
All those who deny that the universe lacks purpose
Forget our role in making the world what it is.

Before I close this chapter, I shall point out that, as long as we remain human, an acceptance of the Grand Unification Theory can not and should not lead to a uniformity of thought. This is reserved for Homo *geneticus*. However, from its Big Picture view the theory reveals the common ground and the relationships that underlie the diversity of our ideas, cultures, races, and ethnicities. The recognition will, inevitable, lead to greater tolerance of differences.

Although idealism—in the form of concern for the lives of others—is an ancient idea, it becomes a new one if we put it into practice every day and patiently inculcate it into our children's lives. Then, if we are sincere, we can say yes to genetic engineering, as we do in the next chapter.

CHAPTER THIRTEEN

GENETIC ENGINEERING? YES.

Gone is the optimism of the Enlightenment that spread from Western Europe to other parts of the world in the 18th century. This period of enthusiasm was also called the Age of Reason. The intellectual advancements of the preceding 17th century fostered the belief that the discoveries of natural laws would be followed by rational solutions to political and social issues. The leaders of the Enlightenment predicted that the state would be an instrument of progress and that science would, in time, solve all human problems.

Unfortunately, as we saw in Chapter Four, the predictions of those halcyon days were faulty but not entirely wrong. We shall touch upon the Age of Reason briefly in the next chapter as well as later in this chapter. It has been wisely said that if we don't learn the lessons history teaches us, the lessons will have to be repeated. Today, we blame almost everything we can think of for the failure of the rosy predictions of the 18th century—too much freedom, too little freedom, too much technology, insufficient technology, lack of acceptance of religion, too much insistence on religion, too much government intervention, not enough government assistance, and so on. Hate-filled people blame persons of another race, ethnicity, nationality, or religion for what is wrong in the world. Others hold those responsible who have a different socio-economic status or political orientation from their own. In the 20th century, accumulated hate finally led to mass exterminations which created greater deterioration of hope and confidence in humankind's ability to solve its problems peacefully.

The problems we must solve are indeed formidable. We encounter corruption in politics, false and misleading claims in the market place, increasing violence and brutality on our streets, and men who seek glory by bombing innocent men, women, and children riding in buses, shopping at malls, or working in their offices. At the beginning of this book we asked: "Is a world where these things happen ready for the application of genetic engineering?" We answered, no. The know-how of genetic engineering can easily spread to nations who, without guilt, have practiced genocide and promoted world-wide terrorism. There was ample justification for entitling Chapter One, "Genetic Engineering? No."

The crucial issue of human health lies closely to the application of genetic engineering. Recently, the March, 1996, issue of *The University of California At Berkeley Wellness Letter,* published by its School of Public Health, issued a warning. The lead article began: "Imagine a world without drug regulations." In the opinion of the editors, "false claims would be made for medicines . . . risks and side effects would go unmentioned." The article continues with predictions that there would be no way to know if pills were contaminated with other substances or whether any pill would contain enough therapeutic ingredients to do what it was supposed to do.

I am not suggesting that gene technology will ever be unregulated. I am certain that every effort will be made to control genetic engineering. Determined government and public bioethics agencies will keep a watchful eye on it. But we cannot assume that regulation will be consistent in all counties of the world. World-wide enforcement of rules and regulations have seldom been successful. Nor has enforcement always been successful in our own country.

The warnings in the *Wellness Letter* targeted the hormone melatonin, but have implications for other substances and procedures as well. Without sufficient evidence, "at least three best-selling books—written by people with M.D.s or Ph.D.s after their names, and blurbed by others with impressive-sounding titles promote melatonin as a natural wonder drug and make breath-taking promises for it." There is no reason to suppose that gene technology will be safe from such exaggerations. In a short time, some aspects of gene therapy will be able to be as easily administered as blood tests are today. Experiments with rats show that gene therapists can efficiently insert new genes into chromosomes with crippled versions of some viruses. Within the next few years, we will have the ability to use genetic diagnostic tests to determine the likelihood that any person will become a victim of any of the inherited diseases.

It is questionable if children who mature to adulthood in our present culture—the one described by Dr. Menninger in the first chapter of this book—will have sufficient moral restraint to handle genetic engineering with enough precautions and confidentiality to insure people's privacy and safety. In one form or another, the scenario of melatonin may be repeated involving some aspect of gene technology. For this reason, a small number of behavioral scientists and other concerned people have reached the conclusion that it would be best for us to delay experiments with our genome until the time arrives when we are committed to a morality that is now generally lacking. In reaching this conclusion, we considered genetic technology as separate from other technologies. In this chapter, we shall look at it as a component of technology as a whole. This will cause us to see genetic engineering in a somewhat different light. Human beings who have the power of abstraction and can conceive of a morality that reaches beyond their own selfish concerns also have the ability to solve their problems and will do so some day, in some way.

We have previously discussed the make-up of reality and the nature of morality. We have searched for priorities and investigated our role in nature. We compared some of the teachings of Jesus with Nietzsche's incisive contentions. We explored freedom of choice and searched for the components of a new mind-set. As we now view gene technology within the larger context of all technology, we are forced, by necessity, to say, with caution, "Yes, go on," because the forward march of genetic engineering can never be halted. Human nature will not permit it. Therefore, it is important that we adopt an ethic that will not require us to follow the well-trod path of Darwinian evolution. That would occur if, in the presence of genetic engineering, natural selection would convert *Homo sapiens* into *Homo geneticus*. Since we prefer to remain who we are, there is hope that this potential danger to our existence as a species may spur us on to save ourselves. The sole purpose of this book is to nourish this hope and perhaps in a small way contribute to its fulfillment—or at least make a start in that direction.

In cases where there are strong temptations to say "yes" and insufficient consideration of consequences, it is far better to say "no" or "wait," as we suggested at the beginning of this book. When we focus on the commonality of gene technology with other technologies that were developed throughout our human history, we discern that a yes answer is possible, and perhaps even unavoidable. Our technologies are a notable characteristic of human nature. I shall write the remainder of this chapter keeping this point of view in mind. We are a can-do species.

Genetic engineering can be seen as a product of a millennia-old technological trend that began when our early ancestors first chipped stones to make tools. Beyond tools and conveniences, technological innovations reflect an irrepressible human urge to explore the unknown and discover how far we can push the outer limits of human achievement. Scientists, as well as of the rest of us, often succumb to the gambler's obsession to beat the odds. The majority of Nobel Prize winners are among those who have done so. Breaking records fascinates us even in trivial matters. Some people were bent on discovering how long a member of our species could perch on top of a flagpole—a craze popular several decades ago. Other attempts to beat the odds attract cheering crowds—such as someone's efforts to remain as long as possible astride a bucking bronco. The hoopla that accompanies such events masks the pervasive, underlying aspect of human nature that is involved. As we establish new records of human endurance, wisdom, and even folly, we may not realize that they are rooted in the dynamics of the universe. Even if they are trivial, they actualize potentialities that shape the universe's ever changing configuration. A thoughtful acceptance of this momentous fact may help us to gain a new impression of ourselves and our unique role as human beings.

Biologists have described nature as innovative, and our tendency to be innovative is reflected in our human nature. Excitement accompanies our technological breakthroughs, ranging from the discovery of the wheel, through heart transplants to gene technology. The Egyptian and Mayan builders of pyramids and the ancient people of Britain who constructed Stonehenge must have experienced the same kind of excitement that today accompanies space travel and genetic engineering. Humans often succumb to the lure of a challenge and overcome their fear of the risks involved. It is noteworthy that we would take a different kind of a risk if we halted research in genetic engineering! I refer to the risk involved in backing away from a technological challenge. Technological projects that require ingenuity help us to avoid boredom and could—if we appreciated them—enhance our collective self-image, instead of creating inappropriate guilt. It is time that we permit our unique role in the universe to help us build up our self-esteem. We would then be more satisfied with ourselves and not so angry at each other. Mutual celebration would lessen the craving for mutual competition. The species and the individual are similar in their need for psychological satisfaction and self-acceptance. Since we can't appreciate ourselves in other ways, winning wars served to promote collective self-esteem. The same scenario holds true for individuals. At least 30 percent of all crimes are committed to gain self-esteem, as well as to provide challenges to young people who are bored.

We tend to undervalue our technological achievements. Contrary to what we may have been told, challenges provided by creating material innovations need not distract us from things that are spiritual. Both may be given their due place in our lives and can coexist. Some material and spiritual things might even be sisters under the skin. Human ability to invent gadgets is awe-inspiring, as archaeologists will recognize if they dig some of them up a few thousand years from now. Unless there were great calamities such as earthquakes that occurred in a few centers of ancient civilizations, devastating epidemics, or destructive wars that left civilizations in ruins, technology has never walked backwards.

A number of years ago, a motion picture, *The Gods Must Be Crazy*, depicted a Bushman (a more correct name is Basarwa) living in the Kalahari desert in Africa who found a discarded Coca-Cola bottle. He thought that the gods must be crazy to leave such a magnificent object behind. The Coca-Cola bottle story was merely a Hollywood invention. Nevertheless, it makes the point that we fail to appreciate our technological abilities. Technological breakthroughs have their dangers but they also help to reveal Big Picture views. That happened when Copernicus refuted the Ptolemaic system of astronomy and proposed that the earth was not the center of the universe. It is happening now that we learn by means of communication technology that we really are one interdependent human race. We could make humans respecting humans as members of the same species a challenge, instead of using racial, religious, and national differences as an easy outlet for accumulated hostilities. This would require that we must first repudiate the theories proposed by the British zoologist, Richard Dawkins, and others who view the "selfish-gene's" role as the standard for finding purpose in human life.

Seen from a Big Picture view, we shall reach a different conclusions from those Dawkins arrived at when he wrote that morality in human behavior should be considered as enlightened self-interest. In the preface to the 1989 edition of his book, *The Selfish Gene*, Dawkins stated that he did not wish to focus on humankind or on the individual person but rather "take the gene's eye view of nature." Instead, I hope that a few of my fellow scientists will spend some of their time to focus on humankind's eye view of the gene than the reverse. I would rather think of the gene as an actualization of the potentiality of the universe than view the universe as an actualization of the potentiality of the gene.

In the mid-20th century, Lecomte du Noüy, a biologist and philosopher, wrote a book entitled, *Human Destiny* (1949). Many people today would agree with what he said. Du Noüy complained that the material side of civilization has conquered "the soul of modern humanity" and keeps people "in a kind of breathless expectation of the next day's manufactured miracle." As a result he found that, "little time was left for solving the true problem: the human problem." By "human problem," he meant the difficulties we have in getting along with one another, the rivalries between nations, the wars, the poverty, and our competitive acquisitiveness. Du Noüy erred by taking the "instead" view of technology rather than the "in-addition-to" view of it. Of course, human concerns are paramount. However, technology need not compete with solving human problems. Quite the contrary is true. We can enlist technology to help solve them. Gene technology, properly used, could greatly increase the world's food supply and could, in time, prevent and cure many devastating diseases. Besides that, every new technological breakthrough represents a new actualization of a potentiality of the universe and deserves respect in that it adds to the universe's reality. This is good as long as it does not harm humankind or the earth which is our home.

My wife is baby-sitting three of our grandchildren 30 miles from where I am writing this. The invention of the telephone enables us to stay in touch with each other and for me to learn how my grandchildren are coping with the temporary absence of their parents. I am wearing glasses that help me continue to read and write. Du Noüy likened our enthusiasm for technological innovations to the fascination of children "in their first view of a three-ring circus." To be sure, a carnival-like media atmosphere seems to accompany all of our major breakthroughs, especially those achieved by molecular biologists. The simple reason for this is that technological breakthroughs are exciting and worth celebrating.

My cautious yes to ongoing research in gene technology is contingent on three requirements. It must not impinge on the right of freedom of choice; it must be conducted by caring people and not solely for profit; it must not be used merely to satisfy curiosity. I will say a few words regarding the last of these requirements. Regardless of consequences, we have not learned and may never learn the lesson of Lot's wife, who was turned into a pillar of salt when she glanced back to view the destruction of the cities of Sodom and Gomorrah. If we succumb to pure curiosity while pursuing genetic therapy, we risk not being turned into pillars of salt but rather becoming members of the species *Homo geneticus*. We share curiosity with many of the animals, but there is a saying drawn from experience: "Curiosity killed the cat." Curiosity has its place in human life, but not when it comes to genetic experimentation with the personalities of human beings.

Cultures, legends, and oral traditions of nonliterate peoples reflect the spirit of human innovation. The urge to redesign human beings, placing a human footstep on the moon, and the descriptions of the heroes (human or animal) of ancient peoples, have much in common. Natives of Meso-America at first welcomed the arrival of the Spaniards riding on horseback and decked out in armor as gods whose coming their legends foretold. Of course, they expected these gods to have advanced technology. Admiring technology is so much a part of human nature that we see it in modern science fiction as well as in ancient legends that describe divinities, spirits, and even medicine men as having the power to create magical technological breakthroughs. If people can't create technological breakthroughs they *imagine* that they can. There is no way out.

Sacred writings enthrall us with their accounts of breakthroughs in "divine technologies" called miracles. Pegasus, in Greek mythology, was a flying horse. Moses drew water from a rock. Buddha walked in the air. Muhammad ascended to heaven on a winged horse. Jesus turned water into wine. The Hindu mystic Ramakrishna conversed with a stone statue. Even today, without claiming to perform miracles, escape artists and stage magicians fascinate their audiences by their seeming ability to perform impossible feats. We even enjoy imagining that a genuine technology exists where it is clearly only an illusion. Therein lies gene technology's hazard and its power. We love to watch magic shows and shall expect genetic engineering to entertain us and do the impossible.

When it does not, people become hostile. I shall mention an example of hostility towards modern technology. After eighteen years of threatening society with mail bombs, the notorious Unabomber was apprehended. He was a mysterious figure who sent his victims letter and package bombs that were responsible for three deaths and twenty-three serious injuries. The Unabomber was clever enough to avoid detection and finally was apprehended only when his brother identified him to the authorities. What was his gripe against society? Its technology!

In a manifesto he sent to the *Washington Post* and the *New York Times* he wrote: "With regard to revolutionary strategy . . . we absolutely insist that the single overriding goal must be the elimination of modern technology." To relieve his boredom, caused in part by his disdain of technology, he found excitement by sending people associated with technology explosives in the mail. Never mind that explosives were among the first technological breakthroughs created in ancient China. By forcing attention on himself, the Unabomber also gained the visibility we discussed in Chapter Four. There we quoted Bishop Berkeley's misleading but influential dictum *esse est percipi*—to be is to be perceived—that still haunts us today. How much better would it have been for the Unabomber and his victims had he used his meager funds to select a few intriguing technological devices from the vast varieties available in mail-order catalogs or in the many electronic sales outlets we find in almost every shopping center. Instead, he utilized the modern technology of the United States Postal Services and that of explosives to express his disapproval of modern technology.

People often buy things that are clever even though their utility is minimal. There is always something thrilling in the thought: "Wow!—this thing works!" We cannot hang our failures on the doorknob of technology any more than we can view technology as the product of our failures. Neither we nor the Unabomber can blame gene technology if, eventually, *Homo geneticus* takes our place as the dominant species. I quote Shakespeare:

> The fault, dear Brutus, is not in our stars,
> But in ourselves, that we are underlings.

If we do not use gene technology in the 21st century exclusively for human good and allow exploiters to manipulate it for personal gain, it will not be their fault, but ours "that we are underlings." Genetic engineering has enabled us to design insect-resistant plants that improve the quality and quantity of our food supply. As I said previously, some think there may be a hidden danger in eating these foods because altering them genetically may result in unpredictable side effects. Others have religious objections to meddling with nature because they view genetic engineering as an affront to the handiwork of their Creator. They believe that changing life forms should not be a human prerogative. We must respect these views.

However, as gene technology increases the world's food supply these objections will be overridden by pragmatism and the successful results of genetic engineering. I do not expect a universal outcry against genetically modifying the bugs, snails, and worms that harbor viruses and bacteria dangerous to humans. Current researchers are working on a method of genetically altering mosquitoes so that they will no longer be able to transmit malaria. An expanding zoo of transgenic animals including mice, pigs, and bacteria—all star performers in genetic engineering—now manufacture human enzymes, proteins, and replacement genes in their bodies. The animals have been involuntarily enlisted to help combat an estimated 3,000 human genetic diseases. Genetically altering animals to reproduce human genes may seem somewhat distasteful but, so far, this use of gene technology has not been seriously opposed. Now on the horizon we see that by modifying some of the functions of master genes that control cell reproduction, we shall eventually be able to cure and even, perhaps, eliminate cancer and other devastating diseases.

As plans for genetic engineering become increasingly more ambitious in preventing diseases and increasing our food supply, it will become correspondingly more difficult to object to its goals. A philosophy of genetics now knocks on the doors of university departments of psychology and sociology, asking to be admitted to disciplines that traditionally championed the importance of environment in shaping behavior. Meanwhile, the argument on the relative influence of environment versus heredity—the old nature/nurture controversy—continues unabated. As I stated at the beginning of this book, I believe that it is essential that environmental techniques take precedence over genetic ones. Psychotherapy must not casually be replaced by gene therapy when matters of personality are involved.

The association between a specific genetic defect and a personality disorder was shown in a study of 104 adults and children from 18 families reported in *The New England Journal of Medicine* (1993). Seventy percent of those with a defective gene were found to have attention deficit hyperactivity disorder. Children with this disorder tend to be inattentive, impulsive, and restless. They experience difficulty in school and social situations because they lack the ability to focus on what is going on around them. The genetic codes for a defective thyroid hormone receptor were identified as being responsible. Since then, more behavioral traits have been linked to a specific gene or a group of genes. It would be unfortunate if a genetic emphasis provided by these studies minimizes important environmental factors that also contribute to shaping behavior.

Recently, a recidivist child molester begged the prison authorities to castrate him. He declared that he was certain that he would continue to molest children when he is released from custody. Castration could be considered a form of genetic engineering, since it excises the organs required for reproduction. The recidivist's request was sensibly denied for reasons I have given in the book—environmental methods, psychotherapy together with medication, should be tried first to treat such conditions. Using genetic engineering to create personality changes is to take the path of least resistance.

We should continue to say no to genetic engineering of human personality traits not directly related to physical or actual mental illnesses except, perhaps, in rare cases of uncontrollable violence or incurable antisocial behavior in which everything had been tried in vain—environment methods, psychotherapy, medications, religious counseling. I have said yes only to those aspects of gene technology that do not involve genetically redesigning the human personality or creating culturally-dependent beauty standards for human appearance.

We have not exhausted the possibilities of helping people in diverse ways by means of innovative environmental intervention. However, no matter how hard we may try to avoid using gene technology to modify personality traits, the danger of gradually branching out in that direction will always remain. Then, once we start redesigning people's personalities by sterilization, castration, or genetic engineering, *Homo geneticus* will be waiting for us around the corner.

In the chapter that follows, we shall try to deal with the question:

"How can we change the world?" In world history, armed conflicts and conquests have changed the world. There are alternatives. Religions have changed the world. Technologies, also, have changed the world significantly—and not always for the worse. We shall explore the role that education may play to help to do it. How well genetic engineering will serve humankind will depend to a significant extent on the kind of education our children receive in the 21st century. I shall offer only a way of begining to change the world by increasing the use of a single word: Why. It is seldom asked anywhere in the world today. In the next chapter I do not offer it as a panacea—only as a first step in how to gain a new mind-set required for the application of genetic engineering.

CHAPTER FOURTEEN

CAN WE CHANGE THE WORLD?

We have tried many times to improve the world, yet nothing seems to have succeeded. History tells us that the Allies justified their sacrifice in World War I, in which a total of 8 ½ million people lost their lives, as an effort "to make the world safe for democracy." Idealism, realism, theology, reason, enlightenment, science, industrialization, existentialism, social Darwinism, modern Darwinism, nationalism, cultism, all failed to bring about a lasting or significant improvement in the way we act towards each other. Instead of acknowledging their ideas as faulty, the proponents of remedies held the people who adopted them responsible for the failures that followed. Today's tragedy may be that we have run out of ideas.

The single course that seems open to us is to now to return to the past and dress up yesterday's failures in today's clothing. We then rename them and call them "new." Darwinian paradigm does this when it offers reciprocal altruism as a solution. In this view, morality is basically selfish but can lead to good results, as when you scratch my back and I scratch yours in return. Eventually, the modern Darwinists claim, no one will need to use a do-it-yourself wooden back scratcher anymore since everyone will be scratching everyone else's back. We will be like apes gooming each other. Then all will be well in the world. If selfishness were the basis of love and altruism, I would have to ask myself how I differed ethically from a mosquito that lovingly alights my skin.

Enough of this! Let us look elsewhere for ways to change the world. On April 20, 1996, there was a news item in the national media about a former Buddhist monk who had been one of Thailand's holiest men—until he was defrocked in the wake of sex scandals and accusations of plagiarism. He was arrested by U.S. authorities for forging immigration documents. Newspapers reported that although the monk was disgraced, he retained thousand of followers throughout the world. Twenty-seven temples in various countries remained loyal to him. In the U.S., the

Buddhist monk gained a large following. What was his magic?

It was not that of a conjurer, although some naive persons believed that he could cure diseases. The paper noted that he taught dharma. In Hinduism this means the ethical duties of virtue and right conduct. In Buddhism dharma refers to righteous truth derived from the teachings of Buddha. His American disciples said that the defrocked monk taught them that cosmic order exists in the universe and that "natural and moral principles apply to all things and beings." His followers said that his teachings answered their questions about why things happen in human life. In their opinions, the monk's teaching applied to both religious and secular questions that no one else had answered.

There was no need for these converts to seek out a defrocked monk to hear about a human/cosmic partnership. The same message that "natural and moral principles apply to all things and beings" can be found in the Judeo-Christian concept of God as One, which clearly implies that His children—the human race—are one. It follows that people of all races, religions, and national origins must be equally accepted and respected. Could it be that this message does not come out clearly in Western religions? To ask "why" leads open-minded people to constructive actions. Questioning why we need doctrinal differences to divide Christians led to the Ecumenical Movement. Significantly, the word ecumenical is derived from the Greek and means, "from the whole world"—the Big Picture view. The Ecumenical Movement attempts to resolve differences in beliefs and create common understanding among Christians. Questioning "why" contributed to other unifying efforts among religions, such as the annual Conference of Christian and Jews. All humans can become citizens of the world without compromising their own religious or national affiliations. There is no disloyalty involved in accepting other members of our species as belonging to the human race.

"Love" is the most important word in human emotions, "hope" in expectations, and "why" is paramount in cognition. No words exist that match these three. Together they make a team that reveals meaning in life. Secular "why" questions lead to understanding and knowledge. Yet questions beginning with "why" are seldom asked in school or at home. Sectarian restrictions are often placed on such questions in places of worship. Serious "why" questions often motivate us to continue to ask "why" even after we have received an initial answer. Think of the trouble I got into after asking, "Why should we continue research in genetic engineering?" The question has led me to struggle with the thorny issues of reality, morality, history, and theology, and it hasn't ended yet. Nevertheless, it is better to live in a "why-do-it?" kind of a world than one only concerned with a "how-to-do-it." This applies particularly to genetic engineering. Many "why" questions encourage us to think deeply about ourselves and others. But caution is required. Unanswered "why" questions can drive some people to strange sources to seek answers—for example, to a defrocked Buddhist monk.

Hedrick Smith, a fellow of the Foreign Policy Institute at John Hopkins School of Advanced International Studies is correct when he writes in Rethinking America (1995): "The first clues to what makes a nation tick—its distinct core value—can be seen in how children are educated. The school is a mirror of society. It reflects the culture of a nation and imprints the same cultural archetype and mind-set on each new generation."

Smith describes the educational policies of Japan and Germany. In Japan, a school child first learns cooperation and community values. After the early primary years, a child is pressured from home as well as by his own ambitions to perform well at school. Japanese students are taught that performance in school leads directly to a job, or to coveted entrance to college.

Smith found that German educators devote a great deal of attention to socializing students when they are young. Class includes team competition, as well as relaxation, and music. As in Japan, parent-teacher cooperation plays a leading role in early German education. Smith concludes that at the elementary level where initial impressions are formed, Japanese and German educators are more successful than their American counterparts in inculcating values and habits of teamwork and high expectation. Nevertheless, it seems, that neither European educators nor those in American schools, encourage students to ask questions beginning with "why"

The word "education" is derived from the Latin word for "to lead out." In almost all educational institutions throughout the world it is practiced as "to push in." Perhaps this helps to account for the fact that not in Japan, Germany or the United States, does education progress from broadening students' sense of local community to world community. Somehow, it does not seem patriotic to do so and anyone who attempts to do it may be suspect. We shy away from the Big Picture view of a united human race. Nevertheless, we must begin to expose students to it to preserve the human genome. Throughout the world, the small picture of local cooperation is stressed because it is the local taxpayers who support the education children receive. We must wait till Christmas to hear the words—"Peace on Earth, Goodwill to all people."

It is in schools that the impact of genetic engineering can make a difference. If children are taught that our human genome is the joint property of all the world's peoples, the thought may help to unite humankind and lead us to have respect for each other. Perhaps it might even foster feelings of community among the different nations.

A united human race represents a deep longing in the hearts of people who are disturbed or made cynical by current events. In the absence of world unity, the sense of community that cults offer attract some people. Cult leaders answer "why" questions that their followers failed to obtain elsewhere. Some of these leaders promise that membership in their cults will provide conditions that will lead to world unity. Thus, they fill a need that creates an almost insane loyalty. In Japan, Shoko Asahara, the bearded , half-blind leader of a cult called, "The Supreme Truth Sect," was responsible for an early morning rush hour nerve gas attack in a subway system that killed and wounded several thousand people. He planned a revolution that was designed to answer questions the Japanese school system failed to address. In Japan, Ashlara's cult is a profoundly troubling social phenomenon, because it attracts brilliant young university graduates.

Certainly, we cannot blame the existence of cults on the educational methods used in schools. However, we may ask: "Could questions that are unanswered elsewhere explain why Asahara's 'Supreme Truth Sect' attracts some bright Japanese university graduates?" Could it also explain why some Americans follow the "righteous truth" teachings of a defrocked Buddhist monk who claims to have answers to Big Picture questions? Neither Japanese, German, nor American curricula put aside time to discuss such questions. The closest I came to that in the schools and universities I attended in the U.S.A. and Europe was during after-school "bull-sessions."

Let us expand on Hedrick Smith's statement quoted earlier in this chapter: "The first clues to what makes a nation tick—its core values . . . can be seen by how children are educated." For the sake of our common genome, children must learn what makes the world, as a whole, tick. To understand "ticking" requires a universal effort to teach children what we all have in common as human beings. This means that we must start to ask important questions beginning with "why" as part of the teaching process. "Why" opens the doors to abstract thinking, which is a unique characteristic of our species. "How" confines answers to particulars. "How-to?" is the joint preoccupation of humans, animals, insects, bacteria, and viruses. Humans alone can ask "why" Thinking "why" could help us resist the temptation to exploit gene technology for personal gain.

"Why" is not a popular question to ask or answer. I admit that such questions can sometimes be outright annoying. It is a humbling experience to ask "why." If honestly asked, it creates awareness that one does not possess all the answers required to live in peace with one's self. It reminds one of the limitations of one's knowledge. This is a lesson many intellectuals of the Age of Reason failed to learn. It explains also why it is often best to encourage persons who asks "why" to try to answer their own questions with the encouragement and resources of a neutral instructor.

In "why" we find the key to the success of cognitive therapy. In my own field of clinical psychology, cognitive therapy has helped clients confront and solve problems. Psychoanalysis is a psychotherapeutic technique that connects the past to the present. A client undergoing this psychotherapy is encouraged to identify residuals of past traumatic experiences. By means of insights, clients obtains relief from repressed memories. A more recent Behavioral Therapy made its appearance on the mental health scene. It helps release a person from previous conditioning that led to automatic problem behaviors, habits, or defensive compulsions. Token rewards and punishment to "recondition" and "extinguish" the problem-causing behaviors are used in this therapy.

In cognitive therapy a carefully thought-out "why" question calls on a client's reasoning power—the unique human part of the person. Cognitive therapy has proven helpful to such an extent that behavioral therapists have incorporated it into their own approach called, behavioral-cognitive therapy. The power of cognitive therapy is that it explores "why" questions with clients with results that enable them to see their problems in a different light and respond to them realistically.

How can we change the world for the better? This is a dangerous question. People aware of their own limitations hesitate to answer it and that, in turn, causes them to avoid thinking about it. They know that no one may ever be able to produce a good answer. Platitudes, propaganda, and too much reliance on only one human attribute, such as rationalism, have failed in the past. False optimism about rationalism occurred during the Age of Reason. Even with the best intentions, unrealistic answers do more harm than good. They have shown us that it is unwise to hold out false hopes. That, too, is the temptation that faces us in the Age of Genetic Engineering, yet "How can we change the world?" remains a question that we continue to ask. Judeo-Christian theology requires the coming of a first (according to Judaism) or a second coming of a Messiah. Eastern religions put their faith in merging with Ultimate Reality or Nirvana.

The 17th and 18th century Age of Reason was discussed in Chapter Four. It was marked by its fervor against something. It could be called the age of controversy because it brought distrust as well as optimism. The movement was promoted by intellectuals who attacked the Church and fought for democracy and humanitarianism. People believed that science could solve all human problems. The Enlightenment was not as much a "why" age as a "how-to" age. Descartes did not ask, "Why am I living?" He asked, "How do I know that I am living?" The leaders of those times asked, "why" in science and produced an overconfident "how" answer to the social application of it. Their promises and predictions failed to occur. This left a deep disillusionment in people's lives that made a revolt against the idealistic ideas of the Age of Reason inevitable.

French writer and philosopher Voltaire (1694–1778) was one of the leaders of the Age of Reason. His morality was based on freedom of thought. Voltaire called upon his contemporaries to act against intolerance and superstition. But he led a campaign summarized in the phrase he often used—"let us crush the infamous ones." He referred to the religions he viewed as harming people. It is regrettable that the Age of Reason had an agenda that proved unworkable. The movement came from the top—the intellectual—down and not from the bottom—the people—up. It bears no resemblance to what we are proposing for a new mind-set that must come the people themselves.

We return to Hedrick Smith's observation, "The school is the mirror of society." "Why" questions today are almost invisible in this mirror.

Newspapers and television screens serve as schools for many adults and they tell us what we should think. In religious education, "why" can be disposed of by, God wants it that way," sometimes followed by a warning—"it is sinful to question Him or Her." In college, "why" is shunted into departments of Philosophy. Some of these classes illuminate students, but more often they do not see the ideas presented by philosophers as relevant to their own lives. This is because, as a rule, students themselves do not participate in searching for the solutions philosophers offer. Students listen, take notes, but seldom ask "why" questions. In some beginning philosophy classes "why" questions suffer a slow death while touring Greece, then sight-seeing in Europe with short side-trips to India and China.

Let us return to the United States. Our educational objectives focus almost exclusively on performance—how to spell, do arithmetic, learn history, properly use English grammar, use computers, and excel in competitive sports. From there we go on our way through life to do things without having a clear idea of why we are doing them. We work hard to answer "how" questions that consist of unconnected fragments adrift from something larger and more significant. "How" is single dimensional. "Why" is multidimensional and leads to other whys as well as to hows, and later, inevitably to more whys and hows. "Why" may never be answered to anyone's satisfaction and in that there is strength, while in certainty there is weakness.

Many teachers prefer not to explore the "why" of things with their students because they themselves were not taught in this way. They may fear that "why" could touch upon values, initiate controversies or, perhaps, cause students to bring up tabooed subjects. The practical reason to avoid "why" questions is that they don't immediately pay off. However, sometimes "why" questions do pay off. In 1993, a 43 year-old postal supervisor from Turlock, California with 25 years of service lost his sight and applied for a disability retirement. The U.S. Postal Service approved the retirement, but a higher review board denied it later on the basis that the postal employee's blindness was caused by a genetic disorder. This was interpreted to mean that his condition predated his employment. As a result, the postal supervisor found that he was not covered by unemployment benefits. A recent study by Stanford and Harvard Medical Schools documented hundreds of cases of such genetic discrimination.

However, persistent "why" questions offered a solution. The postal supervisor kept asking "why" all the way up to the state legislature, which forced the disability provider to admit that genetic predispositions are not the same as a "pre-existing conditions." Similar "why" questions led 26 states to enact laws prohibiting genetic discrimination and other states are now considering doing the same. "Why" is a wonder word. It doesn't always work as well as it did in the postal supervisor's case but "why" has the power to open people's eyes like no other word can. Of course, I refer to those "why" question that are asked sincerely and not those asked in defiance.

If "why" is thoughtfully asked before gene technology is applied, we would be more apt to provide the necessary safeguards. The "why" of applying it to humans would have to explore all of the possible long-term negatives as well the positives, with human benefit and morality as yardsticks. This would have to apply to the commercial use of gene technology as well as to experimentations in laboratories. Then, unlike sterilization used earlier in the 20th century and castration authorized by the courts in 1996, our genome might be safe. However, we cannot expect people who did not ask "why" questions in their homes nor during their school years to ask them when they apply genetic engineering.

Some of today's youths pursue antisocial and unlawful acts because society offers too few alternatives. These young people have never seriously asked themselves any "why" questions. Therefore, they did not discover other ways to live. Some young men and women can lift themselves above their former lives. The few who do so, are able to develop new mind-sets. We hear of their success in jobs and professions. I have questioned some of them about the changes they made in their lives and I received a standard answer. It was something like: "One day I started thinking, 'Why am I doing this to myself?'" This is the question that changed their lives and society.

"Why" led them to find purpose in life. Purpose—called teleology— is an anathema in science. Scientists rightfully avoid assigning purpose when assessing natural phenomena. To imply purpose is a projection that can lead to false conclusions. However, it is possible to view purpose as occurring within Bergson's theory of creative evolution. There we can give human life a purpose simply because our creativity has assigned purpose to it. Purpose is an abstraction, a construct of the human mind, just as morality is. It requires a "why" question to discover purpose. With that discovery, randomness ends as the sole cause of events in the universe. Without an idea of purpose, genetic engineering would become rootless and disconnected. Questions starting with "why" better than any other questions are capable of creating a new mind-set. Nevertheless, as we consider abstractions such as morality and purpose we must learn the pitfalls of illusions that occurred in the Age of Reason. Creative evolution does not lead to utopias.

In the future, as we do now, we shall require rewards and punishment to insure that people will comply with the rules of social behavior. Therefore, it will take a concerted effort by parents, teachers, the clergy, youth leaders, and the entire community to encourage our youths to replace their highly visible "heroes" who perform acts of violence with exciting heroes who struggle to adopt moral values. If given a choice, many of our young people will find values that center on human concerns more appealing than those that use the kind of leverage: "If you don't do what I want, I'll do what you don't want." We can only hope for a moral world in the Age of the Genome if we begin now to give our youth's lives a new sense of direction. Perhaps, if we encouraged them to consider "why" more frequently it would help us make a start. To ask oneself or others "why?" and be unable to obtain a satisfactory answer may be the most informative and enlightening experience of all, for then we might realize that we were on the wrong track or had to obtain more information about what we were proposing to do.

The next Chapter, Conclusions, is a summary of some of the ideas presented in the book.

CHAPTER FIFTEEN

CONCLUSIONS

This chapter presents a review and summary of some of the ideas presented in this book. Beyond that, it offers a number of additional reflections that evolve when the thoughts derived from the individual chapters are combined.

Chapters One and Two began with the conviction that at this point in time we should say no to any further research concerning the application of genetic engineering to human beings. We felt that the pursuit of gene technology requires a morality that would match its unprecedented power to redesign life. We did not see the presence of a guiding hand in our culture today that could safeguard gene technology from exploitation. After we examined reality, morality, human nature, and freedom of choice among other topics, we gave a conditional nod to continuing further research in genetic engineering. Our long journey began on a negative note but ended on a more positive one.

We did not change our mind, but as we proceeded we faced new facts and considered concepts that presented a Big Picture view of genetic engineering. This permitted us to see it from somewhat altered perspectives. We realized that we could not hold back the increasing flow of discoveries and breakthroughs in molecular biology. We had to find a way to live with them and searched for a mind-set which would help us preserve our species while we experiment with the building blocks of life. Chapters One and Two provided ample reasons to call for prudence in this task.

In Chapter Three we offered a brief overview of some of the basics of gene technology. We pointed out that the function of individual genes can be identified by linking them to a disease or characteristic expressed during the life of an organism. A key factor in genetic engineering is that after a gene has been located on the DNA, the expression of the characteristic to which it is linked can be recognized. In addition, it is important to keep in mind that unpredictable factors are always present. For example, there are master genes that switch other genes on or off. Also, exposure to different environmental conditions may activate dormant genes. Therefore, it is more accurate to speak of genetic tendencies than of genetic certainties, although some of these do exist, especially in the area of genetic diseases.

At this time, infections with disabled viruses are the main delivery systems used to reach the areas where replacement of genes must take place. Almost weekly, new discoveries change the information concerning genetic engineering. In this book we did not concentrate on the actual techniques used in gene technology. Many other books are available that fully describe how gene technology works and is applied. Instead, this book, as the reader has learned, devotes itself to an assessment of some of the potential consequences of genetic engineering. At this time, this aspect of gene technology deserves increased attention.

In Chapter Four we reviewed a period of intellectual ferment in the history of Western civilization that helped to shape our present mindset. It was called the Age of Reason and led to democracy, humanitarianism, and fostered hope for humankind. Science flourished and new social ideas emerged. These offered people a hope for a utopian life. However, it placed full faith in human rationalism and failed to take into account the bit of craziness that exists in all of us to which I referred in Chapter Eleven. Brushing this aside led to major miscalculations and did not justify the epoch's title as the Age of Enlightenment.

Some intellectuals who lived during the Enlightenment were skeptics who believed everything had to be demonstrated before it could be accepted as real. This view is incompatible with the totally different definition of reality presented in Chapter Five. In Chapter Four, we mentioned that some of the great figures of the Enlightenment, such as the French philosopher Rousseau, erred in romanticizing the "Noble Savage" as the human ideal. It was typical of the age to view a fragment of reality as the whole of reality, just as today we tend to regard genetic engineering as isolated from other aspects of life. The influential Bishop Berkeley believed that there could be no existence independent of perception. This idea led people to seek visibility in order to prove that they were actually alive and not make-believe persons who did not really exist. The need for personal visibility helped shape the mind-set of the 20th century. It led to the philosophy of Existentialism, which added the concept of existential anxiety, in a world already trembling with insecurity.

Even today, many people view themselves as the source of their own souls. Neurobiologists ascribe the origins of our mental equipment to the construction of our brain and take it for granted that we are miraculous potentialities of the universe. In Chapter Five, I mentioned that our human potentialities do not originate within the contents of our skulls. It's the other way around. The content of our skulls merely actualize and manifest these potentialities.

Reality is the yardstick which we use to establish the validity of all things. In Chapter Five we found it to be an unreliable yardstick. On further probing, we viewed reality as created by all happenings. Therefore it is an unconvincing yardstick with which to measure things, including the value of genetic engineering. Reality breaks easily into fragments that can then be taken to represent the whole of reality. This leads to short-sighted conclusions. One example given in the book was that an error exists when one thinks that a morality that serves a useful purpose is a genuine morality. Morality cannot serve any purpose other than to exist.

While we are dealing with the subject of reality in Chapter Five, we add here that it has wide implications for both life and death. The reason why genetic engineering will be involved in the subject of death is because there are genes that have been identified as "death genes." It is correct to say the days of our lives are numbered, not only by circumstances but more surely by the genetic clock ticking within our "death genes." Altering the timing of these genes or turning them off could, theoretically, extend our lives or even give us immortality. When in the 21st century gene technology is able to prolong human life, theologians, politicians, physicians, attorneys, and the general public will be embroiled in a controversy even more contentious than the present one that deals with abortion.

In Chapter Five, we made the point that our lives have meaning because everything we do and think contributes to eternal reality. The laws of nature do not permit the existence of nothingness which some equate with death. Over a hundred years ago, a Chief of the Suquamish Indian Tribe, Seattle Seath, put this thought into one sentence. "There is no death, only a change of worlds." This sentence elegantly summarizes what I attempted to say in Chapter Five. However, there is one difference. In Chapter Five, we added the thought that we live in the "other world" from the moment we draw our first breath. The idea is important, since the perceptions of life and death will play a significant role in the future application of gene technology.

To do justice to the subject we shall take ideas from several chapters in the book. In ancient times, craftsmen were buried with their tools so that they could resume their work in their afterlife. Some rulers were buried with riches, food, and symbols of their power, as well as their wives and members of their court, so that they could continue to rule in the next world. "We shall not cease to exist" is expressed by countless monuments erected all over the world. They are echoed by ancient mummies waiting patiently to be awakened. When they were alive, they failed to view their thoughts and acts in life as eternal monuments of the universe. And thus they did not realize that their death was not the end, but rather a "move to the different world" in which they already lived while their hearts were still beating on earth.

Let us just carry this thought a little further. In Chapter Five we suggested that we live three lives concurrently: 1) the life we recognize as ours, 2) an eternal life created by our actualized potentialities, and 3) our lives as catalysts for actualizing other potentialities. Since the potentialities of the universe we actualize give us immortality, we need no grave marker, nor an afterlife, nor an immortality gene inserted in our DNA. If we can accept this concept, we will not have to wait to be rewarded in an afterlife. We are rewarded each day we live by the fact that our lives have helped (in a small way, to be sure, but nevertheless helped) to shape the universe into what we want it to be. The same thing may be said of cows and cockroaches. But we can do it in our unique human way that includes, morality, compassion, and justice— things that are important to us. We can spend part of our lives making our eternal immortality one of which we can justly be proud. This is our eternal monument.

The utilitarian doctrine holds that the value of anything is determined solely by how we might benefit from it. Mill had that in mind when he viewed the greatest happiness for the greatest number of people as the only legitimate use of morality. In that case, the use of genetic technology to create happy people would be the most efficient means for achieving this goal. In Chapter Six I stressed the opposite, that human innovation consists of our ability to value a thing for itself. What would become of romance if lovers adopted the view that when they love someone they are merely disguising a selfish goal initiated by their genes instead of their hearts—which is what modern Darwinists would have us believe.

The subject matter in Chapter Six—morality—will be the key to the beneficial use of gene technology. In Chapter Six we express our view that humans possess a capacity that is unique in nature. It is fitting that Chapter Seven centers on human/animal comparisons. As we examine Chapter Seven we find that we, alone of all living forms, will be able to create our own evolution by design. This is no reason for human chauvinism or for feeling superior to other creatures that inhabit the earth. We cannot and do not wish to claim superiority for humankind. Animals are superior to us in many important ways including the crucial one— the capacity to survive. We point out only that we are unique. One might say we march to the beat of a different drummer. The French philosopher, Henri Bergson called this drummer, "creative evolution." Therefore we are not totally at the mercy of Darwin's theory of natural selection. With this in mind, we continue on to Chapter Eight to evaluate our priorities.

It is not amazing that, in concert with all other species, our highest priority is to survive. What is surprising, however, is the reason we wish to survive. It is not merely the survival of our species but, as I have continually insisted, to mine the treasury of the unactualized potentialities of the universe or, to put it into a different context, to reveal God in all His Glory.

Our strength is not found in clever adaptations, sharp claws and canine teeth, or in life-saving appendages like wings or in any special physical agility. Our strength is in our ability to live in a different world of abstractions where, in our imagination, we can actually do things that the universe itself is powerless to accomplish. If we would stop and think about this we would realize that this ability, though confined to our imagination, is nevertheless stupendous! We can defy gravitation, travel faster than the speed of light, design perpetual motion machines that counter the Second Law of Thermodynamics. The human imagination has contributed a prize jewel to the treasury of the cosmos. Throughout the book, we asked the reader to recognize this endowment and not allow it to slip through his or her fingers into the waiting hands of a

Homo geneticus. It could occur if we outsmart ourselves by genetically redesigning ourselves to become him or her.

We confront the question of our role in nature in Chapter Nine. We decided to do it within the context of harmony and struggle. The chapter asks: "Could we tolerate harmony, since it would be boring beyond belief?" God in His mercy endowed Eve with an irresistible curiosity and the naiveté to serve as the instrument that freed humankind from Eden's chains and enabled us to discover a world of potentialities. What loving Parent would otherwise expel His children and all of their descendants from home merely because He found their hands in the forbidden cookie jar? Loving parents ask the children to leave home to attend an out-of-town college, as did Adam and Eve—figuratively—after their expulsion. Banishment from Eden was not punishment, but liberation! Jesus did not die for our sins, but for our liberation. And that makes His sacrifice immensely more meaningful. Too long a sojourn in Eden would have effectively dulled our minds. Nevertheless, we continue to long for harmony more than for anything else. Freudian psychoanalysts might equate it with an unconscious wish to return to the womb.

Humans used their post-Eden wisdom and knowledge to find a way to eat their cake and have it too. The answer to the riddle of how we did this is revealed in Chapter Nine. Simply, it is that we constantly struggle to create and maintain harmony. This gives us the best of two worlds —harmony and struggle both. It offers us Heaven on earth with occasional visits to Hell for contrast and comparison.

Continuing in this vein, we called attention in Chapter Nine to the cruelty of the food chain whose basic rule is "eat or be eaten." Several chapters in the book mentioned the unique human attributes that lead to guilt. Merely to continue to live requires us to be "sinners" from the point of view of those who see only harmony in nature. Only humans have laws that require the merciful slaughter of animals. In some of their legends, ancient hunters describe their prey as willingly offering their lives to become food for hunters. This is a pathetic way to avoid guilt for our membership in the food chain club. Many native people have food taboos that suggest guilt. Guilt is reserved for humans—in nature, only the victims of the food chain seem to voice disapproval by their vocalizations of fear and pain. Hebrew law goes one step further than the ancient hunters did. It forbids Jews to consume calf meat and drink its mother's milk at the same meal—a clear token of guilt. The writers of the Judeo-Christian Bible reflect their discomfiture with participation in the food chain by prophesying the arrival of an age in which the lion and the lamb will lie side by side in peace.

Let us dwell on this thought just a little longer. Vegetarian animals patronize the food-chain restaurant as well as predators do. Plants are living things with life's universal genetic blueprint and have some genes that could be interchanged with some of ours. Many plants have developed specialized techniques designed to protect themselves from destruction—thorns, skin irritating chemicals, foul odors, and powerful poisons. Even if grown solely for the purpose of providing food, plants may not be happy to die. Had they the genes required to inform us of this they would do so. However, it is killing mammals, which cry out in pain or are frozen into silent terror when killed for food, that evokes human guilt. Let us note—our human guilt actualizes a most peculiar potentiality of the universe.

In Chapter Ten, we continue to speculate on matters theological so unique to humankind. Jesus is described as placing forgiveness above retribution. This calls for transcendence. Transcendence can be achieved by moving beyond the small picture view of a happening to a Big Picture perception of it. Sometimes we find that we can transcend negative feelings if we view an event within its larger framework. Our first impression of it may have distorted its meaning. The parable of the Prodigal Son referred to in Chapter Ten illustrates this.

Recently, on Easter Sunday, 1996, a crowd of people followed a man wearing a crown of thorns on his head and carrying a large cross as he walked towards Broadway in downtown San Diego. The crowd sang the moving hymn, "Were you there when they crucified my Lord?" A minister addressed and told them, "Jesus continues to carry the cross, through the boardrooms and hospitals, fields and factories." He asked, "Do we ever notice Him there, or are we too distracted to by the business at hand?" It seems that we can be so distracted by the presence of Nietzsche's supermen who occupy these places that we do not pay attention to anyone else.

The new Darwnists, like Wright (1994), attempt to squeeze the teachings of Jesus into the philosophy of Nietzsche to make them both compatible with natural selection. Wright cites John Stuart Mill (1863) who wrote, "In the Golden Rule of Jesus of Nazareth, we read the complete spirit of the ethics of utility. To do as one would be done by, and to love one's neighbor as oneself, constitutes the ideal perfection of utilitarian morality."

However, a different interpretation of Jesus' teaching is possible than one based on the utility of mutual personal gain. It is that the references to oneself were meant to convey identification with another human being. If we identify with others, we experience their feelings. In this way we would be able to gain sympathy and compassion.

Nietzsche and the modern Darwinists describe how we often act but Jesus revealed to us how we could act and what we could become if we adopted his mind-set. Nietzsche would view "turning the other cheek" as a sign of submission after a struggle for dominance as other animals do. Jesus, in contrast, taught us to recognize it as a sign of dominance over oneself—a much greater accomplishment.

Freedom of choice is the concern we addressed in Chapter Eleven. There we found that freedom is a mixed blessing, in that it entails responsibility and obligation. At the time of the writing of this chapter, the newspapers carried articles that reported controversies about whether or not to respect the wishes of persons who are mentally competent and wish to end their lives when they suffer from painful terminal illnesses. Organizations for and against this freedom of choice have been formed. Bronowski's definition in Chapter Six suggests that it is immoral to prolong persons' lives by using medical methods against their wishes. Today, dying patients are victims of play-gods. When it is time to die, only play-gods—who do not understand the meaning of morality—would attempt to deprive terminally ill persons of their right to die with dignity.

Let us look at gene therapy from the view of freedom of choice, discussed in Chapter Eleven. In contrast to changes that are created by genetic engineering, biochemical personality changes created by prescribed drugs can be reversed. They require cognitive participation by the patient, even if only to the extent of permitting an injection or taking a pill. This act, though a minor one, nevertheless represents the essential element of self-involvement and control. Freedom of choice could not be passed on to future generations who inherit the genetically designed make-up of their parents.

We suggested from the beginning of this book that a new mind-set is necessary to insure the safety of our genome. In presenting this in

Chapter Twelve, we determined that it is not a mind-set that has been derived from Darwinian evolution. Darwinian evolution has created a mind-set reflected in Social Darwinism. Although we disavow it today, we still conduct our lives according to its tenets. A new mind-set must rely on a model that envisions purpose in human life beyond that of the "selfish gene." We can find it in the perception of a Grand Unification Theory, described in Chapter Twelve. If we accept this theory and apply it, we shall find purpose in our lives..

In Chapter Thirteen we view gene technology within the larger framework of human technology and recognize it as human enterprise that began almost with the beginning of our species and helped us to become what we are. Technology has been attacked and has become a popular scapegoat as a surrender to material values. Technology began with the chipped flint that made a primitive tool. We must not make it a scapegoat for our problems. Some people feel guilty when they obtain pleasure from things "not of the spirit." They should remember how much the empty Coca-Cola bottle impressed the fictitious motion picture Bushman who assumed it was a work of the gods. All things that reflect human ingenuity actualize potentialities of the universe and to use them constructively is fun—not sin. Confused moralists fail to understand that enjoying clever modern inventions does not represent succumbing to material values. They are actualized potentialities in common with pyramids and cathedrals.

From this point of view, gene technology finds its place within other technologies that helped us to create self-actualization. We cannot avoid our attempts to better the lives of humankind by our discoveries and genetic engineering is no exception.

Chapter Fourteen brings us to the seminal question, "Can We Change The World?" Let us ignore the apparent grandiosity of the question and recall what we said about reality in Chapter Five. Then we can rephrase the question and ask: "Are we able to *avoid* changing the world?" In terms of the philosophy we have presented, the response is clearly, "No!" A more helpful question might be: "Can we change the world to make it a safe place to apply genetic engineering?" My final answer is yes—after we have taken sufficient time to explore "Why shall we do it?" instead of hastily asking, "How shall we do it?"

We suggested in Chapter Fourteen that we might begin to create a new mind-set by encouraging students, from an early age, to ask "why" more often. This seems a big task for a very small word. Will and Ariel Durant, authors of the monumental series, *The Story of Civilization*, wrote their final impressions in the book, *The Lessons Of History*, (1968). These are as pertinent today as when they were written: "Consider education not as a painful accumulation of facts and dates and reigns, nor merely necessary preparation of an individual to earn his keep in the world, but as the transmission of our mental, moral, technical, and aesthetic heritage as fully as possible to as many as possible, for the enlargement of man's understanding, control, embellishment, and enjoyment of life."

Chapter Fourteen gives us a formula to achieve this end. As admitted, it may be oversimplified, but is a good start. It consists of gearing the educational system to encourage questions that begin with Why instead of only with How. As noted in Chapter Fourteen, "Why" opens our minds while "How" closes them immediately after we find the answer. "Why?" has no merit if it is motivated by hostility instead of in the spirit of exploring an idea to gain understanding. The ability to understand is one of the components of creative evolution within which genetic engineering must eventually find its place.

Fortunately, we shall have a powerful new ally in the near future. I do not refer to a Messiah from Heaven—but, surprisingly, to what Nietzsche called, the "Demon within us." It is the drive to power that supermen would use to control the world by means of gene technology. These would convert us into the species *Homo geneticus*. Fear of being converted into *Homo geneticus* may force us, finally, to become a united human race. Unfortunately, love by itself has not changed the world. Love combined with fear may have a better chance of doing so. I use "fear" in the sense of "The fear of the Lord is the beginning of wisdom." (Psalm 111:10). That wisdom includes awe, reverence, and respect for life as it now exists. Let us never forget that we have the ability to overcome because we are a can-do species, genetically engineered that way by nature and aided by the values we adopt. These values must be derived from the Big Picture view of the universe—the Grand Unification Theory described in Chapter Twelve. It reveals our relatedness to all that exists and creates a respect for all there is. If we use this or any similar theory to guide us, we need not fear where genetic engineering will lead us. Again I refer to the wisdom of a Native American—a Sioux holy man, Black Elk, (1863–1950),

Hear me, four quarters of the world—
A relative am I!
Give me the strength to walk the soft earth,
A relative to all there is
Give me the eyes to see
And the strength to understand . . .

The thrust of the universe to actualize its potentialities is the original invisible hand that creates reality. Scientists sometimes see only the secondary manifestations of this hand's work and make the error of viewing them as origins. This has created wide-spread disenchantment with science, such as that noted by Janet Raloff who stated in Science News (Jan. 1986, p. 360), "A large and growing share of the population rejects aspects of science." I agree with Raloff, who sees this as regrettable. She observes, "rejection of scientific truths and logic is . . . eroding support for teaching critical thinking. . . ." Many people prefer to look for a Big Picture view and reject a jumble of pieces and parts of knowledge thrown together under the roof of academia. Philosophy and theology, in different ways, pull things together and give them meaning. It is, therefore, important, that we do not respond to genetic engineering as a fragment of reality but place it within a Big Picture.

The signs of the times indicate we are moving exactly into the opposite direction. An article in the Journal of the American Psychological Society, *Observer*, (May/June, 1996) suggests that evolution should be taught to psychology students. Under the heading of "Teaching Tips" this article is titled, "Using Evolution by Natural Selection as an Integrative Theme in Psychology Courses." In the June, 1996, issue of the same journal, an article argued that natural selection explained why stepchildren are at greater risk of becoming victims of violence than genetic children. The reaches the generalization: ". . . it is especially in the domain of social motives that psychology needs Darwinism. . . ."

Behavioral scientists, as others, often find what they are looking for. Had the authors searched beyond natural selection, they may have considered that stepparents are more likely to become irritated by stepchildren simply because they have no role in producing them, just as we are more annoyed by the noise others create than our own. Stepparents are often forced to accept a stepchild unwillingly as the price paid for the marriage bargain. This makes them especially sensitive to the affection their mates display on their genetic children in contrast to themselves. It naturally raises the suspicion, "did you marry me to impose on me a surrogate parental obligation for your child's sake?"

In my own practice of psychology, I have noted that a sibling rivalry between a stepparent and a child often develops. Regardless of the animal comparisons, among human, this is a more likely explanation than that a stepparents "selfish gene," driven by natural selection, mindlessly creates a scenario of violence.

We have not learned our lesson. Darwinian evolution applied to humans has a Freudian flavor with only weak underpinnings that time will reveal. It would be more enlightening for students to rediscover psychology's grounding in philosophy than learn "critical thinking by exposure to evolutionary perspectives" in which all human activity is seen as self-serving. At least Freud postulated a censoring Superego, the civilizing influence of the environment on human behavior that could account for the origin and development of a creative evolution. On the other hand, natural selection is confined to the Freudian Id—instincts and desires. It is restricted only by the Ego, the self-serving modifier of the Id.

Throughout the book, I use the word universe as some readers in a different context would refer to God as a divinity. As a scientist, I tried to remain as much as possible within the realm of science. The words in this book may be placed into any context acceptable to the reader. At times, I have anthropomorphized and described the universe in a way that makes it appear alive. I shall borrow a phrase from William James, the American philosopher and psychologist, which explains why I felt the need to do so.

In *The Varieties of Religious Experience*, (1958), James wrote "The universe is not a mere It to us but a Thou, forced on us we know not whence. . . ." This is the only perception of the universe from the Big Picture view that can help us answer the question: "Genetic Engineering—Yes, No or Maybe?"

INTERPRETATIVE GLOSSARY

Note: An asterisk (*) indicates that the word or phrase has special or an extended meaning.

Actualizations of the potentialities of the universe * — They flow through everything that exists and produce an automatic continuation of events triggered by previous events. Nothing can escape involvement in actualizations, since they create reality. Self-actualization, described as a psychological need, is an actualization of the potentiality of the universe that trickles down into the life of an individual person.

Age of the genome — Includes the period that genetic engineering plays a major role in human lives.

Animal/human differences — Biologically, they consist of the structure of our human foot associated with our upright gait, the relative size and complexity of our brain that leads to our capabilities, the long period of human infancy and early childhood dependency that promote nurturing and learning and our capacity for language. We are unique in our ability to make abstractions that reveal new, exciting potentialities of the universe. Abstractions lead to our emancipation from total dependency on Darwinian evolution.

Anthropomorphize — To attribute human attributes to non-human things and events. The reverse, to dehumanize, robs people of their human qualities and reduces them to the level of things. They can then be mistreated without guilt.

Antibiotic — One of a group of organic compounds that can destroy bacteria and inhibit the growth of microörganisms. However, an increasing number of strains of bacteria are becoming immune to these compounds by mutating. This poses a threat to human life and a major medical challenge to genetic technology.

Antisemitism — Prejudice and discrimination against Jews. The use of generalizations and distortions in order to blame Jews for wrongs for which they are not responsible. Blaming Jewish people as an ethnic or religious group as a scapegoat. Sometimes antisemitism is rooted in an emotional inability to apply Jesus' teaching of tolerance and love. Antisemitism rejects Jesus' own ethnicity. Symbolically, it represents a second crucifixion of Jesus since it obliterates the values Jesus died for.

Bacteria — One-celled microorganisms which multiply by simple division. From the point of view of adapting they are among the most successful life forms. Some cause disease while others are useful to both humans and animals. They will become even more valuable to us when they are bioengineered to serve human needs. See *Antibiotic.*

Big Bang — The primal explosion (or time-warp) that scientists believe started the universe an estimated 12 to 15 billion years ago. The theory has been debated for various reasons and so has the implication of its name. However, the term is still in common use among cosmologists to signify the beginning of our present universe.

Biotechnical — The application of technology to biological processes for industrial, agricultural, and medical uses.

Can-do* — May be viewed as the human effort to give a directional focus to the dynamics of the universe with the intention to overcome obstacles.

Chaos — The occurrence of an event that causes a succeeding series of events in ways that do not follow the rules of traditional statistics. It appears as random, erratic, and unpredictable. A new science has evolved from the study of chaos that uses a nonlinear approach to probe for order and patterns in chaotic events.

Chromosomes — One of a group of thread-like bundles of different lengths and shapes that contain the genes which make up the genome of a species. Their number varies among different species. Chromosomes usually occur in pairs. Most cells in the human body normally contain 23 pairs, giving our genome a total of 46 chromosomes that, together, hold an estimated 100,000 or more genes.

Cults — See *Involvement*.

Deoxyribonucleic acid — See *DNA*.

DNA — Initials that stand for deoxyribonucleic acid. Mainly found in chromosomes that contain the hereditary information of organisms.

*Dynamics of the universe** — Forces rooted in the nature of the universe that permeate everything and account for the "push" that causes things to happen. These, in turn, create reality. Life gives them a directional focus and human life can redirect the focus by transcending. See *Transcend*.

Enzyme — A compound that serves as a catalyst in biochemical reactions. Enzymes are complex proteins that create change in the function of other substances without changing themselves.

Eugenics — A theory which proposed that the human race would be improved by restricting propagation among those considered genetically unfit. It aimed to encourage breeding between individuals with genetic traits considered desirable such as intelligence, health, and leadership. It could become a forerunner of genetically redesigning humans.

Eukaryote — An organism whose DNA is enclosed by membranes to form a nucleus within a cell. In contrast, the DNA of prokaryotes lies free in the cytoplasm. Eukaryotes are characteristics of plants and animals; prokaryotes evolved first and comprise mainly of bacteria.

Evil — Sin, vice, wrong doing, offer us freedom of choice. Without evil, goodness would have no meaning.

Evolution — The theory of natural selection. Natural selection is not the only determining factor. In nature, sexual selection by females attracted to male courting displays and colors, plays a role. If human females sexually selected males possessing the capacity for genuine morality, in time any contributing genes would be dispersed throughout our species. It might be a long wait but it would be for good.

Evolution, creative — The theory of human selection. Some genes transmit behavioral tendencies instead of certainties. Environmental conditions influence the expression of their characteristics. This enables humans to use their imagination and make abstractions that create environments which do not draw on the expression of certain genes. Thus we can initiate Bergson's creative evolution with selected environments which circumvent natural selection's self-centered goals and these environments enable a genuine morality to exist. See *Morality*.

Existentialism — A philosophical movement that developed in Western Europe in the 20th century. It maintains that there is no fixed human nature. Humans must have free expression rather than conform to given rules. Its emphasis is on responsibility for one's own actions. Existentialists point out that this could lead to feelings of helplessness and despair — "existential anxiety."

Freedom of choice — Permits maximum self-actualization, but must be limited by accountability and should not inhibit the freedom of choice of others. Without these constraints, it becomes dangerous and destructive.

Gene — The basic unit of hereditary material located on a chromosome that, by itself or with other genes, determines a characteristic of an organism.

Gene, dominant — In humans, carries a trait in a person's genetic make-up that is expressed in the life of the individual. See *Gene, recessive*.

Gene, recessive — Carries a trait that is not expressed in an individual's life unless paired with another recessive gene carrying the same characteristic. It is passed on to offsprings who will not exhibit its trait if it is paired with a dominant gene that lacks it. Recessive genes are more apt to carry undesirable characteristics including genetically caused diseases.

Gene, selfish — When applied to humans, the idea of selfish genes produces a false description of purpose. It represents the smallest picture view of the Biggest Picture ever created.

Gene therapy — Consists of repairing or replacing damaged, faulty, genes that cause disease. Ethical questions would arise if it were used to prevent personality disorders or to correct genetic tendencies considered disruptive to society.

Genetic engineering — Also called recombinant DNA technology. It involves the direct introduction of foreign genes into an organism's genetic material by micro-manipulation at the cell level. It enables correcting defective genes as well as gene transfers between widely different forms of life and may change the inherent nature of an organism.

Genome — A complete set of chromosomes carried by each cell of an organism. The human genome contains all of the genetic characteristics that individuals will develop over their lifetimes. Mapping a genome enables geneticists to alter different forms of life including our own. See *Chromosome*.

Grand Unification Theory (GUT)* — In physics, refers to the unification of the four forces of nature. I have borrowed the idea to suggest a similar theory that includes everything that exists in the universe. It attempts to bring human beings into a Big Picture of a unified cosmos.

*Happenings** — Actualized potentialities of the universe.

*Harmony and struggle** — A theoretical state of tranquillity, balance, and quiescence often longed for. If we could attained it permanently, we might not be able to tolerate it without losing our unique human characteristics. Struggle is present whenever there is challenge. Struggle need not involve violence or strife. It will occur naturally in the minds of people who see both sides of any issue. We may experience a feeling of harmony by accepting struggle as an aspect of life and making peace with it.

Heliocentric earth — The idea that the earth is the center of the universe and everything spins around it.

Homo geneticus* — A term I coined to represent an imaginary genetically engineered species having programmed characteristics. Theoretically, it could replace us if we redesign our species by means of genetic engineering.

Homo habilis — can be translated as "the handyman" or as I have used it, our "can-do" ancestor. Some dating places this group of hominids as having lived from 1.2 to 1.8 million years ago. Their remains are found with crude stone tools fashioned by the use of other modified tools. See *Tool.*

Homo sapiens sapiens — Our human species. *Sapiens* is repeated to differentiate our present species from the preceding one. The term *Homo sapiens* is more commonly used.

Hormone — A chemical messenger released by a certain type of gland and transported in the blood to a specific organ. It acts to stimulate growth, metabolism, sexual reproduction, and other body processes.

How?* — The word is aimed at solutions. Today, we are inundated by solutions that become tomorrow's questions. We need more questions starting with "why" to avoid the small picture world created by answers to questions that begin with "how." See *Why?*

Human nature* — Involves more than merely behaviors originating in upper and lower brain centers. Human nature is shaped by heredity, environment, and the dynamics of the universe that have led to evolution and creative evolution.

Imagination* — Human imagination is the most astonishing of all actualizations of the universe. Within its framework, it provides the mind with unlimited power. Imagination has preceded or accompanied all great human achievements and played a major role in the evolvement of creative evolution.

Involvement — Stems from the need to actualize the potentialities of the universe. It is required to achieve self-actualization. Wisely chosen, involvement represents one of the finest of human attributes. Sometimes, however, it can create fanaticism proportionate to the pointlessness of the goal of the involvement. An example may be found in cults. Often, the greater the irrationality of an involvement, the more zealously it is defended. Self-righteousness serves to deny the lack of an involvement's justification for existing. Wars and conflicts can result.

Jesus* — As an historical figure who taught that people have the power to transcend greed, selfishness, and avarice. If his teachings were put into practice, they would create good will, happiness, and satisfactions. If more people would accept Jesus as their role model there would be no racial hate and less reason for concern that gene technology would be exploited for selfish gain in the 21st century because Jesus placed power over self above power over others.

Logical order, reversal of* — Takes happenings out of their logical sequence, or out of their place in a priority of values. It leads to the defeat of one's purpose as when one puts the cart before the horse.

Leisure — Can create an opportunity for a change in one's usual method of attaining self-actualization or it can lead to boredom. *Homo geneticus* will not have the capacity to experience boredom. Boredom results from the lack of opportunity or know-how for self-actualization, and often can lead to antisocial behavior, crime, violence, prejudice, hate, wars, or fanatic espousal of causes that lack merit.

Manifesting* — To show or show off what one actualized. In this book, I have contrasted manifesting with actualizing the potentialities of the universe. On the positive side, manifesting occurs in communicating, teaching, sharing, and informing. Neutrally, it may be used to gain acceptance and reassurance. On the negative side, its purpose may be to call attention to one's accomplishments for the sake of vanity or because one desires additional status and recognition.

Meaning in life* — Is found in living it, because by living we create reality. We cannot avoid giving life meaning by whatever we do or even don't do. Those who actualize the universe's potentialities *selectively* give their lives a special meaning.

Mind-set — Thoughts that lead to actions. The contemporary mind-set does not provide for the safety of genetic engineering, since primarily it consists of thoughts centered on "what's in it for me?" To make gene technology a safe enterprise, we must turn our thoughts, from early childhood on, to "what's in it for humankind, as a whole, as well as for me?" See *Morality, Why?*

Morality, false or pseudo* — "Morality" that results from genetic programming, indoctrination, fear, or serves as a means to an end.

Morality, genuine* — Results from a voluntary commitment to an ethical way of life in which there is consideration for others. In such a life, one usually finds value placed on honesty, fairness, self-control, duty, and dedication. The overriding criterion of morality is that it is freely chosen without ulterior motives.

Nietzsche—Champion of the power of natural selection and biological evolution in contrast to Jesus' unpopular prescription of humility which draws on creative evolution. See evolution, creative.

Neocortex — The part of the thin, gray outer layer of the brain's cortex usually associated with human thought and higher intelligence.
Neuroanatomy — The anatomy that deals with the nervous system and the brain.

Nucleotides — Comprise the mosaic that helps determine the functions of the genes. Nucleotides may be viewed as the building blocks of the DNA and RNA.

Nucleus — An organelle of plant or animal cells containing the genetic information and controlling the cell's activities. Organelles are subcellular structures with a particular function. The largest organelle is the nucleus. Organelles allow division of labor within the cell.

Parts, more than sum of* — People who recognize it view human life as more than a fragment of reality. This is because, in one way or another, we continually search for our relationship to the whole. See *Spiritual*.

Protein — One of a large number of substances that are important in the structure and function of all living organisms.

Purpose — May be seen as the directional thrust given to the dynamics of the universe by humans who seek self-actualization. Purpose is derived from creative evolution and can be contrasted with randomness, chance, and opportunism.

Reality* — Everything that is or was actualized since the beginning of the universe contributes to the creation of reality. Facts and fiction represent different kinds of reality and should not be confused or replaced by each other.

Recombinant DNA — DNA formed by crossing over genetic material from one form of life to another. It is sometimes used as another term for genetic engineering.

Sex* — Sex involves the flow of the dynamism of the universe through living organisms. The sex instinct pushes life to actualize reproduction. Humans who take a Big Picture view of sex consider sex-with-love as a more advanced contribution to the nature of the universe than sex prompted solely by instinctual drives. This is because sex-with-love actualizes two potentialities of the universe simultaneously—a biological one and one that involves abstractions. Together they lead to further actualizations, for example, poetry and song, loyalty, dedication and sacrifice.

Slave mentality* — People who are greatly influenced by the media could be seen as having slave mentalities or, as Nietzsche might have put it, they have the minds of a "herd of sheep." Naive people are the most vulnerable to all kinds of propaganda and, therefore, most likely to become future victims of gene technology.

Social Darwinism — a view of society that accepts the fundamentals of biological evolution as, for example, the survival of the fittest. Social Darwinists believe that competition without restraints of any kind would establish morality. Darwinists consider themselves as realists and view idealism and humanitarianism as leading to social deterioration. Creative evolution is beyond their range of vision.

Spiritual* — Average-sized dictionaries give up to sixteen different definitions for the word. Therefore, its meaning in a specific context is often unclear. Most definitions describe spiritual as consisting of qualities that are awe-inspiring and lift people's minds above material values. Spiritual things consist of more than that which is obvious and apparent. A relationship of an individual to the Big Picture view may be conceived of as a spiritual one since it brings reverence and appreciation. Things that are spiritual create values, meaning, and ideals that transcend self-centered concerns. Spirituality is beyond the sum of the physical and material parts that make up a human being.

Struggle — See *Harmony.*

Supermen — The German word does not have a male gender and is incorrectly translated. *Mensch* is the German word for human being. Supermen should read, "super human beings" or superpersons.

Tool — A device that facilitates can-do. Apes and some other animals use tools occasionally, but the later *Homo habilis* went one long step further. He and she discovered that an object they fashioned into a tool could, in turn, be used to modify other objects that became more complex tools.

Transcend —To go above or beyond something else, resulting in a profound change of view. In religion transcend sometimes is used as being in touch with God. In meditation it could mean arriving at an altered form of consciousness. Transcending usually causes a change in emotions and motivation. It may reveal a Big Picture view of an event previously seen only with personal involvement.

Transgenic — Pertains to an organism that has received genes from a different kind of organism.

Universe* — Used in this book as composed of everything that has happened in the span of its existence. Beyond that, it is the source of dormant potentialities that have not yet, and may never, become actual ized. One might view these as opportunity. The dynamic flow that runs through all things pushes them to actualize the potentialities of the universe. More than anything else on earth humans respond to that push. See *Actualization, Involvement.*

Viruses — Extremely small infectious particles that can only reproduce in living cells. Outside of these cells viruses are inactive. They have a core of DNA or RNA surrounded by a protein coat. Viruses use the energy of their host's cell to replicate themselves. When they have been genetically disabled, they may be used as vehicles to transport genes to the DNA of specific tissue and, thereby, they play an important role in genetic engineering.

Visibility* — Need for visibility as used in this book does not have its roots in a desire for power, vanity, nor represent a symbolic way of spreading one's genes. Without denying other origins, it represents, specifically, a fear of not really existing unless observed by others. This anxiety causes one to place more weight on gaining recognition (visibility) for having accomplished something than on actually accomplishing it. The result is deleterious to society because it tends to make constructive values secondary to recognition. See *Logical order, Mindset.*

Why?* — If this is asked to consider the causes of and reasons for events, the question should be asked constructively, i.e., not in deficance or mockery. For children it would open up the window to the Big Picture view of life. If these kinds of "Why?" questions were encouraged at home and in the schools, the person who asks them should first be given the chance to work out an answer for him or herself. Then the questioner should be able to obtain ideas from more knowledgeable persons who may openly disagree with each other but are receptive to new ideas.

BIBLIOGRAPHY

Ackermann, R. J. *Nietzsche: A Frenzied Look*. Amherst, University of Massachusetts, 1990.

Bahm, A. J. *The World's Living Religions*. New York, Dell, 1964.

Barash, D. P. *Sociology and Behavior*. New York, Elsevier, 1977.

Barrow, J. D. and Tipler, F. F. *The Anthropic Cosmological Principle*. Clarendon Press, Oxford, 1986.

Beachcrist, L. *Science News*. July 22, 1995.

Bellamy, E. *Looking Backward 2000–1888*. Boston, 1889.

Bergson, H. *Creative Evolution*. New York, Holt, 1911.

Birx, H. J. *Man's Place in the Universe*. Arcade, NY, Tri-County, 1977.

Bronowski, J. *Science and Human Values*. New York, Harper & Row, 1965.

Cohen, J. and Steward, I. "Our Genes Aren't Us," *Discover*, April, 1994.

Colton, C. *The Treasury of Religious Spiritual Quotations*. New York, Stonesong Press, 1994

Conover, C. E. *Moral Education in Family, School, and Church*. Philadelphia, Westminster, 1962.

Corballis, M. C. *The Lopsided Ape: Evolution of the Generative Mind*. New York, Oxford, 1993.

Crick, F. H. C. *The Astonishing Hypothesis: The Scientific Search for the Soul*. New York, Scribner, 1994.

——*What Mad Pursuit: A Personal View of Scientific Discovery*. New York, Basic, 1988.

Crick, F. H. C. and J. D. Watson, "The Double Helix." *Nature*. April 25, 1953.

Cross, C. *Who Was Jesus?* New York, Atheneum, 1970.

Cullmann, O., Tr. Gareth Putnam, *Jesus and the Revolutionaries*. New York, Harper & Row, 1970.

Culotta, E. *Science*, Vol. 265, July 15, 1994.

Darwin, C. *The Origin of Species*. London, J. Murray, 1859.

—— *The Descent of Man*. London, J. Murray, 1871.

Dawkins, R. *The Selfish Gene.* New York, Oxford, 1989.

Diderot, D. *Pensée sur l interpretation de la nature.* 1754.

Drummond, H. *The Ascent of Man.* New York, Burt, 1894.

du Noüy, L. *Human Destiny.* New York, NAL, 1949.

Durant, W. and Durant, A. *The Lessons of History.* New York, MJF Books, 1968.

Eliot, T. S. *Murder in the Cathedral.* San Diego, Harcourt, 1935.

Frye, C. *A Yard of Sun.* Oxford, Oxford, 1979.

Galton, Sir F. *Hereditary Genius.* 1869.

Goodfield, J. *Playing God: Genetic Engineering and the Manipulation of Life.* New York, Random, 1977.

Goodspeed, E. J. *The Life of Jesus.* New York, Harper, 1950.

Gould, S. J. *Ever Since Darwin.* New York, W. W. Norton, 1977.

Greunbaum, A. "Time, Irreversible Processes, and the Physical Status of Becoming," *The World of Physics.* J. H. Weaver, Ed., New York, S&S, 1987.

Halacy, D. S. *Genetic Revolution, Shaping Life for Tomorrow.* New York, Harper, 1974.

Hardy, S.Vignettes *Pitfalls of Evolution.* Science. 263:1301.

Hardy, S. *Review Reinventing the Future—Conversations with the World's Leading Scientists, a Vignette in Science*, 263:696. Also see Harrisson, E., 1994

Harrison, E. *Masks of the Universe.* New York, Macmillan, 1985.

Hawking, S. W. *A Brief History of Time.* New York, Bantam, 1988.

Hopfield, J. J. Review of "An Envisioning of Consciousness," by F. H. C. Crick, *Science,* 263: 696, 1994.

Huxley, A. *Brave New World.* New York, Harper. 1932.

James, W. *The Varieties of Religious Experience.* New York, NAL, 1958.

Jasper, K. *The Way to Wisdom, An Introduction to Philosophy.* New Haven, Yale, 1954.

Kauffman, S. A. *The Origins of Order.* New York, Oxford, 1993.

Kaufmann, W., Ed. *The Portable Nietzsche.* New York, Viking, 1976.

Keddie, N. R. "Comparative Reflections on Hindu Extremism." *Contention.* Vol 2. No.3, Spring, 1993.

Kelly, W. L. and A. Tallon. *Readings In The Philosophy of Man.* New York, McGraw-Hill, 1967.

Kent, S., Ed. *Farmers as Hunters—The Implications of Sedentism.* Cambridge, Cambridge, 1989.

Kent, T. C. *Conflict Resolution, A Study of Applied Psychophilosophy.* Woodbridge, CT, Ox Bow, 1986.

Kent, T. C. *Poems for Living.* San Diego, Human Science Center Press, 1995.

Lee, T. F. *The Human Genome Project: Cracking The Genetic Code of Life.* New York, Plenum, 1991.

Mann, C. C. "War of Words Continues in Violence Research." *Science,* 263: 1375, 1994.

May, R., Ed. *Existence.* New York, Basic, 1959.

Mayr, E. *The Growth of Biological Thought.* Cambridge, Belknap, 1982.

McGee, G. *Center for Bioethics Newsletter*, p. 5, Vol. 1, No. 5. Spring 1996.

Meir, J. P. *The Marginal Jew.* New York, Bantam, 1991.

Menninger, K. *Love Against Hate.* New York, Harcourt Brace, 1942.

Mill, J. S. and Bentham, J. *Utilitarianism and other Essays.* New York, Penguin, 1987.

Mitchell, S. *The Gospel According To Jesus.* New York, Harper, 1991.

Moffat, A. S. "Microbial Mining Boosts the Environment, Bottom Line." *Science*, 1994.

Murdock, D. *Niels Bohr: Philosophy of Physics.* Cambridge University Press, Cambridge, 1987.

Native American Wisdom. Running Press, Philadelphia, 1994.

Nietzsche, F. W., Tr. W. Kaufmann. "The AntiChrist" and "Thus Spoke Zarathustra." *The Portable Nietzsche.* New York, Viking, 1976.

O'Malley, W. J. "The Moral Practice of Jesus." *America.* April 23, 1994.

Pendick, D. "Science & Society." *Science News*, 143:22, May 29, 1993.

Raloff, J. "When Science and Beliefs Collide." *Science News*, Vol. 149, No. 23 June 8, 1996.

Roth, J. K. *Freedom and the Moral Life.* Westminster Press, Philadelphia, 1969.

Schneider, S. H. and P. J. Boston, Eds. *Scientists on Gaia.* Cambridge, MIT, 1992.

Schweitzer, A. *The Quest for the Historical Jesus.* New York, Macmillan, 1968.

Shannon, P. "Gaia Without Mysticism." *Skeptical Inquirer.* Fall, 17:1, 1992.

Shreeve, J. "Lucy, Crucial Early Human Ancestor, Finally Gets A Head." *Science,* 264:34, 1994.

Smith, H. *Rethinking America.* New York, Random House, 1995.

Taubes, G. "Heisenberg's Heirs Exploit Loopholes in His Law." *Science,* 262: 1376, 1994.

Teilhard de Chardin, P. *The Phenomenon of Man.* New York, Harper, 1961.

Tipler, F. J. *Physics and Immortality.* New York, Doubleday, 1994.

Trungpa, C. *The Myth of Freedom.* Boston, Shambhala, 1987.

Van Den Berg, J.H. *The Changing Nature of Man.* New York, Dell, 1961.

Warnock, M. *Existentialist Ethics.* Oxford, Oxford University Press, 1978.

Watson, J. D. *The Double Helix.* New York, Dutton, 1986.

Wilson, E. O. *Sociobiology.* Cambridge, Belknap, 1980.

Wingerson, L. *Mapping Our Genes.* New York, Dutton, 1990.

Wright, R. *The Moral Animal.* New York, Pantheon, 1994.

Index

Name Index